An Atlas of the
CLINICAL MICROBIOLOGY
OF INFECTIOUS DISEASES

Volume 1

Bacterial Agents

THE ENCYCLOPEDIA OF VISUAL MEDICINE SERIES

An Atlas of the
CLINICAL MICROBIOLOGY OF INFECTIOUS DISEASES
Volume 1

Bacterial Agents

Edward J. Bottone, PhD, Diplomate ABMM

Mount Sinai School of Medicine
New York

Foreword by

Gary V. Doern, PhD

Section Director, Clinical Microbiology
University of Iowa College of Medicine, Iowa

The Parthenon Publishing Group
International Publishers in Medicine, Science & Technology

A CRC PRESS COMPANY
BOCA RATON   LONDON   NEW YORK   WASHINGTON, D.C.

Published in the USA by
The Parthenon Publishing Group
345 Park Avenue South, 10th Floor
New York
NY 10010
USA

Published in the UK and Europe by
The Parthenon Publishing Group
23–25 Blades Court
Deodar Road
London SW15 2NU
UK

Copyright © 2004 The Parthenon Publishing Group

Library of Congress Cataloging-in-Publication Data
Data available on application   2002025346

British Library Cataloguing in Publication Data
Bottone, Edward J.
   An atlas of the clinical microbiology of infectious diseases
   Vol. 1: Bacterial agents
   1. Diagnostic microbiology - Atlases
   I. Title
   616'.01'0222

ISBN 1-84214-219-4

First published in 2004
No part of this book may be reproduced in any form without permission from the publishers
except for the quotation of brief passages for the purposes of review

Composition by The Parthenon Publishing Group
Printed and bound by T.G. Hostench S.A., Spain

Contents

This Atlas is dedicated to the many mentors, colleagues, and students who, upon entering my life, paused in transit, left their indelible imprimatur, and enriched its passage. For the past 42 years, my journey has been cradled and embraced by the unwavering love and devotion of my wife IdaMarie. Without her constant inspiration and clarity of vision, this effort and life's true essence would be without meaning. My daughters Laura and Rina, son-in-law Robert and grandchildren Lauren, Vincent, and Camille continue to illuminate my days with the radiance of their being. Lastly, to my parents Lili and Salvatore, although deceased, for the humanity that, through poverty and perseverance, distinguished their lives and transcends time.

Foreword

In this text, *An Atlas of the Clinical Microbiology of Infectious Diseases*, Volume 1, *Bacterial Agents*, Dr Edward J. Bottone has done practicing clinical microbiologists a huge service. Drawing on over 35 years of experience, he has compiled an extraordinary collection of photos and photomicrographs pertaining to the laboratory diagnosis of a vast number of different bacterial infectious disease problems. The material is presented in a coherent and logical manner. The quality of the images in this Atlas is exceptional. As an added feature, pursuant to each pathogen, Dr Bottone provides text which clearly and concisely delineates important features related to cellular and colony morphology, growth patterns, microbiological characteristics, pathogenesis and disease manifestations.

The practice of clinical microbiology during the past 20 years has undergone remarkable change. What was once strictly a manual endeavor, more art than science, has increasingly embraced various forms of instrumentation in the day-to-day provision of diagnostic services. This trend will certainly continue as we enter the exciting new age of molecular infectious disease diagnosis. Unfortunately, largely as a consequence of increased reliance on instrumentation and other forms of technology, expertise in more traditional aspects of clinical microbiology, such as use of cellular and colony morphology as means for presumptive if not definitive microorganism identification, has clearly eroded. This reality is problematic in so far as morphology remains the cornerstone of the laboratory diagnosis of infectious diseases. *An Atlas of the Clinical Microbiology of Infectious Diseases* represents an elegant reminder of this reality.

This text will quickly become an invaluable and practical resource for individuals working in clinical microbiology laboratories. *An Atlas of the Clinical Microbiology of Infectious Diseases* is one of those books that will not collect dust on the shelf; it will be referred to constantly in the laboratory. In addition, training programs for medical technologists, pathology residents and infectious disease fellows will undoubtedly lean heavily on this text as an important tool for teaching. Dr Bottone is to be commended on this wonderful addition to the clinical microbiology literature.

Gary V. Doern, PhD

September, 2003

Gram-positive, branching and non-branching, non-acid-fast, bacillary species

ACTINOMYCES

Actinomyces are Gram-positive filamentous bacteria which are part of the resident oral flora, colonizing especially the gingival crevices and the tonsils in the absence of clinical disease. *Actinomyces* may also colonize the vagina and gastrointestinal tract. Actinomycosis is a chronic disease caused by several species of *Actinomyces*, the most common of which are *A. israelii*, *A. gerencseriae*, *A. naeslundi*, *A. viscosus*, *A. odontolyticus* and *A. meyeri*. Actinomycosis develops when the microorganism is introduced into surrounding tissue and is characterized by the formation of suppurative abscesses that usually result in draining sinuses. Because of radial growth of the bacterium, colonies form in tissue, surrounded by a marked inflammatory response which gives rise to firm yellowish granules ('sulfur granules'). These may be extruded with draining pus. While three major categories of infection (cervicofacial, thoracic, abdominal) are well recognized, actinomycosis may develop in any organ or body site, e.g. the eye (lacrimal canaliculitis) and the female pelvis in association with the presence of an intrauterine contraceptive device. Actinomycosis may occur in conjunction with other bacterial species such as *Actinobacillus actinomycetemcomitans* and *Eikenella corrodens*.

Morphology

Actinomyces ('ray fungus') species are Gram-positive bacteria that occur as branching, beaded, filamentous rods which fragment into short 'diphtheroid' and coccoid forms. Often, the filaments of *A. israelii* have a swollen, clubbed terminus.

Culture characteristics

Actinomyces species are microaerophilic to anaerobic. The optimum temperature for growth is 37°C. With the exception of *A. viscosus* and *A. neuii*, *Actinomyces* species are catalase-negative. *Proprionibacterium acnes*, however, a Gram-positive anaerobic bacterium which resides on the human skin, and which also forms branching bacillary forms, is also catalase-positive and must be differentiated from *A. viscosus* and *A. neuii*.

BACILLUS SPECIES

Members of the genus *Bacillus* are Gram-positive to Gram-variable, aerobic, rod-shaped bacteria which

Figure 1   *Actinomyces*   Gram stain of cervicofacial abscess drainage showing beaded filaments and filament fragments, many of which are slightly curved with clubbed ends bearing a morphological resemblance to *Corynebacterium* ('diphtheroids') species

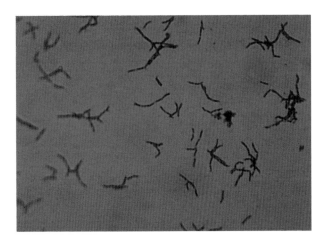

Figure 2   *Actinomyces*   Gram stain of 48-h agar culture showing predominance of branching, beaded filaments with rounded or clubbed ends

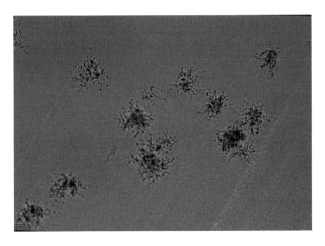

Figure 5   *Actinomyces*   'Spider' colony as viewed microscopically (1000x) after 48-h incubation showing radiating, branched filaments with clubbing emanating from a central inoculum

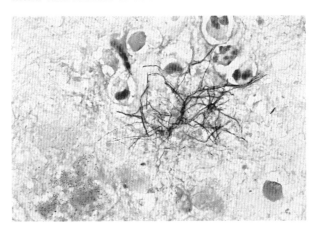

Figure 3   *Actinomyces*   Radiating, intertwining, beaded filaments in Gram stain of pus from brain abscess. Note (lower left) presence of small Gram-negative coccobacilli morphologically consistent with *Actinobacillus actinomycetemcomitans*

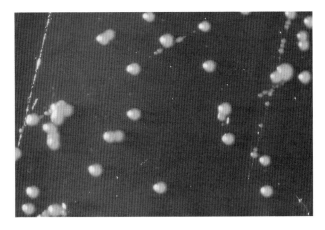

Figure 6   *Actinomyces*   Mature colony of *A. israelii* with irregular contours and 'molar tooth' central depression

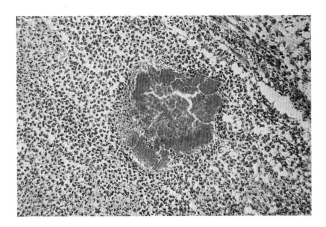

Figure 4   *Actinomyces*   Characteristic 'sulfur granule' in hematoxylin and eosin stain of lung biopsy. Granule comprised of dense filamentous bacterial mass surrounded by prodigious inflammatory response, which together can inhibit antibiotic penetration into granule (colony) center

Figure 7   *Actinomyces*   Discrete granular growth 1 cm beneath thioglycolate broth surface. Each granule represents a single colony

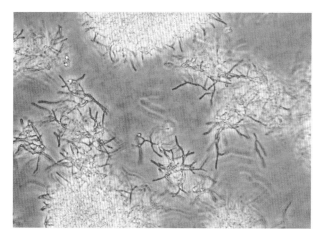

Figure 8 *Actinomyces* Microscopic (1000x) observation of crushed granule from broth culture showing ramifying filaments analogous to those seen in crushed sulfur granule

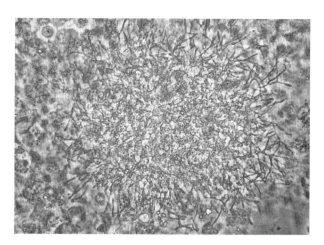

Figure 11 *Actinomyces* Ramifying filaments with clubbed termini in phase-contrast-enhanced examination of wet preparation of crushed sulfur granule extruded into gauze dressing of patient with thoracic actinomycosis. Note the presence of inflammatory cells

Figure 9 *Actinomyces* Extruded washed 'sulfur granules' from chest wall drainage

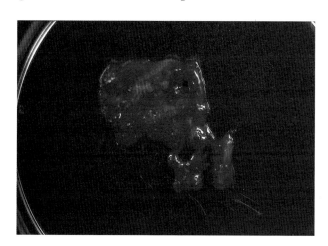

Figure 10 *Actinomyces* 'Sulfur' granules in gauze dressing containing purulent drainage from patient with thoracic actinomycosis

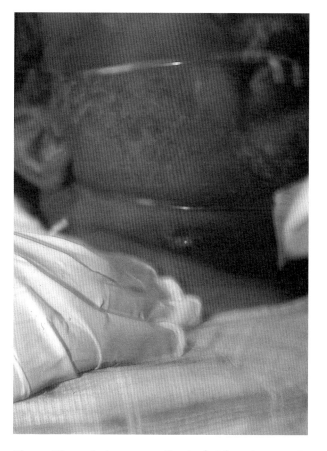

Figure 12 *Actinomyces* Cervicofacial actinomycosis with draining sinus tract

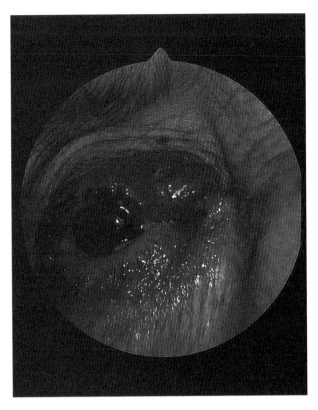

Figure 13 *Actinomyces* Lacrimal duct canaliculitis in an 82-year-old patient. Note the enlarged lacrimal punctum due to actinomycotic blockage and inflammation. Gentle pressing on the punctum may express a 'sulfur granule'

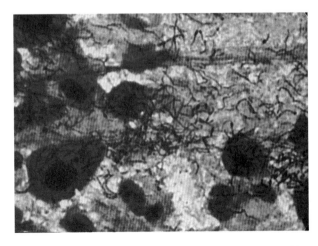

Figure 14 *Actinomyces* Tangled mass of branching filaments, some with clubbed ends, in direct smear of purulent material expressed from lacrimal punctum

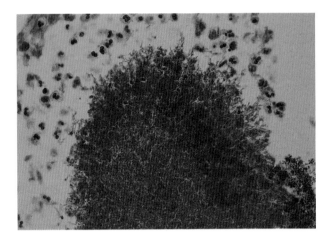

Figure 15 *Actinomyces* Pseudosulfur granule from oral mucosa comprised of densely packed oral flora about a central foreign body (food, plaque). May be differentiated from *Actinomyces* sulfur granule by Gram stain showing diverse bacterial flora with granule periphery, composed of streaming Gram-negative fusobacteria and Gram-variable *Leptothrix* bacilli

form endospores and may be motile or non-motile (*B. anthracis*). They are widely distributed in nature. *B. anthracis*, the causative agent of anthrax, is the only bona fide human and animal pathogen. *B. cereus* is pathogenic when introduced, e.g. by trauma, into human tissue and some strains produce two exotoxins, giving rise to either a diarrheal syndrome, or emetic disease subsequent to toxin ingestion in contaminated foods, especially fried rice. Anthrax, a disease dating back to antiquity, is primarily a disease of herbivores transmitted to humans by contact with spores from infected herbivores. Anthrax may present as a cutaneous infection from contact with contaminated material in which spores enter through abraded skin, as an intestinal infection from consumption of spore-infected meat, or as a pulmonary infection from inhalation of spores. Virulence factors of *B. anthracis* include its polypeptide capsule and a three-component toxin (protective antigen, edema factor, lethal factor) encoded by two plasmids designated pX01 and pX02, respectively. *B. cereus* produces a number of tissue-destructive exoenzymes, including two distinct hemolysins, one of which, cereolysin, is a potent necrotic and lethal toxin. *B. cereus* (predominant), as well as other *Bacillus* species, has been implicated in a number of infections including bacteremia, wound and burn infections, endophthalmitis, meningitis and osteomyelitis. Many *Bacillus* species produce antibiotics such as bacitracin and polymyxin.

Morphology

Bacillus species show a range of morphologic forms, from straight bacilli with parallel or slightly curved sides, with slightly rounded or square ends, to bacilli singly or in long chains. Spores are present which may vary in shape from oval to slightly elongated and appear anywhere along the bacillary tract. Some spores bulge into the bacillary body. On Gram stain, spores are seen as colorless oval bodies along the bacillus. In direct smears from infected sources (skin, lung), *B. anthracis* occurs as encapsulated rods in chains. *B. anthracis* is non-motile.

Culture characteristics

Bacillus species grow on a variety of culture media. On 5% sheep blood agar, *B. anthracis* produces non-hemolytic, dry, white to gray colonies with a striated surface. Most, but not all, other *Bacillus* species produce β-hemolytic colonies.

CORYNEBACTERIUM

Corynebacterium species are Gram-positive, short to medium length bacilli with a propensity for dividing

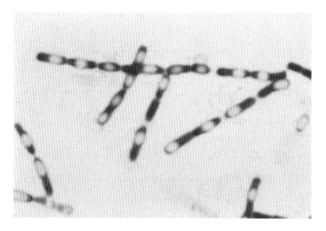

Figure 16 *Bacillus anthracis* Gram stain of culture showing typical filamentous bacilli with square (boxcar) ends and unstained spores

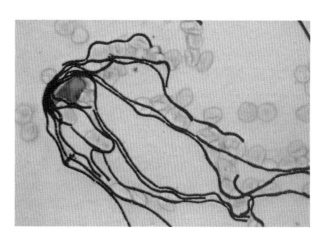

Figure 18 *Bacillus anthracis* Long bacillary forms in chains in blood culture of patient with pulmonary anthrax (courtesy of Tom Robin)

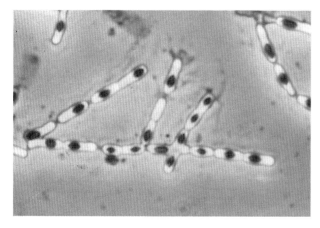

Figure 17 *Bacillus anthracis* Phase contrast-enhanced Gram stain of agar culture. Bacilli with square ends and single, spherical to ellipsoidal, subterminal to equatorial spore slightly bulging into the bacillary body

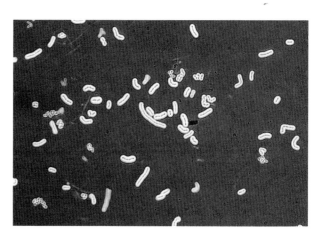

Figure 19 *Bacillus species* India ink preparation showing capsules similar to those observed with *B. anthracis*

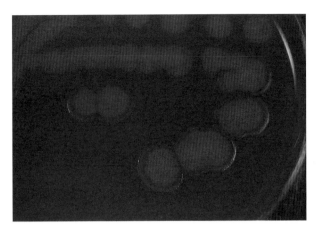

Figure 20   *Bacillus anthracis*   Non-hemolytic, flat, grayish colonies on 5% sheep blood agar

Figure 23   *Bacillus cereus*   Methenamine silver stain of brain biopsy of patient who developed fulminant hemorrhagic necrosis of brain subsequent to intrathecal induction chemotherapy. Note the large numbers of bacillary forms in necrotic areas. *B. cereus* was isolated from several blood cultures

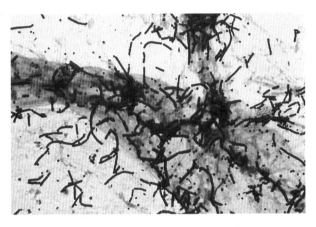

Figure 21   *Bacillus cereus*   Gram stain of eye aspirate from patient who developed endophthalmitis 12 h after cataract surgery, showing innumerable bacilli singly and in short chains. Wet preparation of aspirate showed motile bacilli

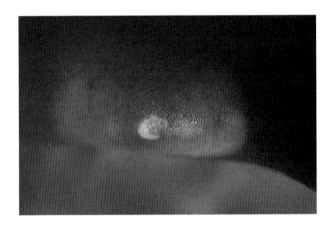

Figure 24   *Bacillus cereus*   Necrotic   lesion   with surrounding cellulitis at site of intravenous line

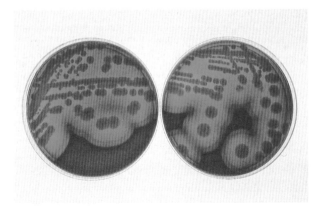

Figure 22   *Bacillus cereus*   Blood agar culture of eye aspirate showing large double zones of β-hemolysis. *B. cereus* hemolysin (cereolysin) is highly tissue-destructive

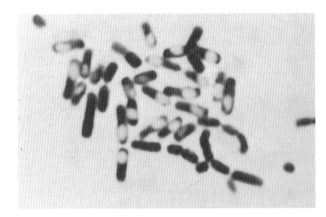

Figure 25   *Bacillus subtilis*   Large bacilli with sub-terminal spores in smear of agar culture

bacilli to remain adherent one to another prior to separation, thereby forming stacks of bacilli (palisade formation). They are widely distributed in animal and environmental sources. While more than ten *Corynebacterium* species have been identified, only one, *C. diphtheriae*, the causative agent of diphtheria, is a well-recognized human pathogen. Toxigenic strains of *C. diphtheriae* secrete a phage-encoded toxin at their site of replication (nasopharynx, wound) which, when taken up by cells, inhibits protein synthesis by inactivating elongation factor 2, leading to tissue necrosis. In the absence of neutralizing antibody, toxin enters the bloodstream and is transported to target organs, such as the heart, kidneys, peripheral nerves, and the brain. In the nasopharynx, toxin-mediated necrosis leads to the formation of a tough, adherent pseudomembrane consisting of dead host cells, polymorphonuclear leukocytes, and fibrin. In severe cases, respiratory tract obstruction may occur. Susceptibility to *C. diphtheriae* toxin is indicated by a positive Schick test. The organism is acquired from human carriers by inhalation of infected droplets.

Other *Corynebacterium* species such as *C. jeikeium* cause infections infrequently and only under specific host conditions such as neutropenia and immunosuppression. *Arcanobacterium* (*Corynebacterium*) *haemolyticum*, a β-hemolytic, catalase-negative, colonizer of the skin and nasopharynx, occasionally causes pharyngitis similar to that caused by group A streptococci (*Streptococcus pyogenes*).

Morphology

Corynebacteria are highly pleomorphic, ranging from slender, slightly curved bacilli with rounded swollen ends, to short ovoid bacilli resembling cocci. *C. diphtheriae* contains dark-staining, reddish-purple inclusions (metachromatic granules) when stained with methylene blue. Corynebacteria are non-motile, non-sporulating, and non-encapsulated. *C. aquaticum* is motile but its classification as a *Corynebacterium* is tentative.

Culture characteristics

Corynebacteria, which are facultative anaerobes, grow well aerobically on a variety of bacteriologic media. For *C. diphtheriae*, two primary enrichment, selective and differential media are recommended. Loeffler's coagulated serum slant is used for the primary isolation from nose and throat cultures. On this medium, *C. diphtheriae* grows luxuriantly, forming gray to white colonies. Because serum is present, *C. diphtheriae* colonies may pit the medium surface due to the action of proteolytic enzymes. Liquefaction of the entire medium will occur over time. Tinsdale blood agar is a selective and differential medium containing potassium tellurite, which inhibits normal oropharyngeal flora, bovine serum, and cystine. *C. diphtheriae*, as well as several other *Corynebacterium* species, produces black or brownish colonies (tellurite reduction) surrounded by a brown halo due to proteolytic activity. *Corynebacterium* species are catalase-positive.

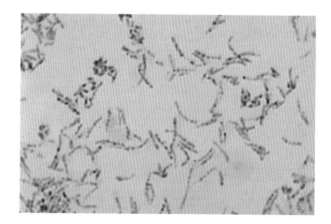

Figure 26 *Corynebacterium diphtheriae* Gram stain of culture showing pleomorphic, slender, slightly curved, unevenly stained bacilli with distinct dark-staining metachromatic (Babes–Ernst) granules

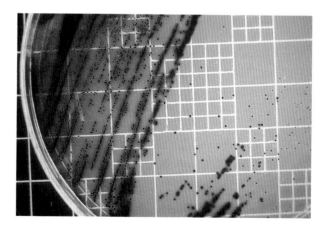

Figure 27 *Corynebacterium diphtheriae* Small black colonies on Tinsdale agar due to reduction of potassium tellurite in medium to tellurium, a black substance, which is incorporated into the colonies. Courtesy of J. Michael Janda, PhD

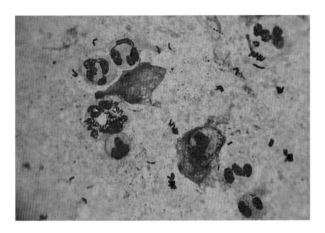

Figure 28 *Corynebacterium jeikeium* Gram stain of peritoneal fluid from a patient who was undergoing ambulatory peritoneal dialysis and who developed peritonitis. Note the presence of slightly curved bacilli singly, in palisade formation, and intracellularly in polymorphonuclear leukocyte

Figure 29 *Corynebacterium jeikeium* Short, almost coccal, clubbed bacilli arranged singly, and in palisade formation. Smear from agar culture

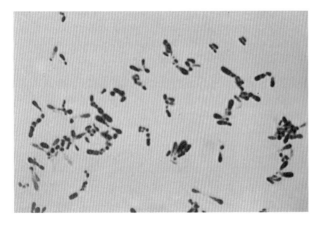

Figure 30 *Corynebacterium xerosis* Short, slightly curved 'diphtheroid' bacilli with clubbed ends arranged in short stacks (palisade formation)

CORYNEBACTERIUM JEIKEIUM

This lipid-requiring cutaneous coryneform causes serious local and systemic infection in immuno-compromised patients, especially those with neutropenia. The species is characterized by multiple antibiotic resistance except to vancomycin. It is a major nosocomial pathogen. *Corynebacterium urealyticum* is a urease-positive, lipophilic, multiple antibiotic-resistant species which causes urinary tract infections primarily in association with alkali-encrusted cystitis. *C. urealyticum* bacteremia occurs mainly in elderly long-term hospitalized patients who have been urologically manipulated.

ERYSIPELOTHRIX RHUSIOPATHIAE

Erysipelothrix species are pleomorphic Gram-positive, non-spore-forming, non-motile, aerobic bacilli found among swine, mammals, poultry, and fish. Swine are a natural reservoir for the microorganism and, during episodes of bacteremia, large numbers of the organism are shed in their urine and feces. *E. rhusiopathiae* causes mainly cutaneous infections among individuals who have occupational exposure (butchers, abattoir workers, fish handlers) to the urine, feces, saliva, and tissues of infected animals. Access is through breaks in the skin. Septicemia may occur in the absence of cutaneous infection and may lead to endocarditis. Virulence of *E. rhusiopathiae* is related to capsule formation *in vivo* and concurrent resistance to phagocytosis. Intracellular survival in macrophages in non-immune hosts also enhances virulence.

Morphology

In culture, *E. rhusiopathiae* may occur in a smooth or rough colony form accompanied by a corresponding morphologic presentation. Smooth colony forms show uniform morphology of small, slightly curved rods singly or in small groupings, whereas, in rough colonies, long filaments predominate.

Culture characteristics

Erysipelothrix grows poorly on most bacteriologic media. On 5% sheep blood agar, colonies are pinpoint after 24-h incubation. At 48 h, colonies may be surrounded by a zone of greenish discoloration (α-hemolysis) and two discrete morphotypes may be observed. In triple sugar iron

Figure 31 *Erysipelothrix rhusiopathiae* Gram stain from smooth-colony morphotype showing uniformly staining slender, slightly curved bacilli with rounded ends

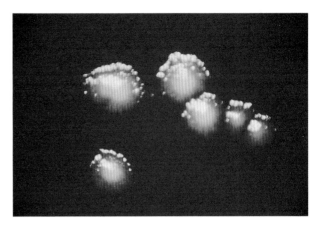

Figure 34 *Erysipelothrix rhusiopathiae* High-power magnification of rough colony depicting uneven fimbriate border

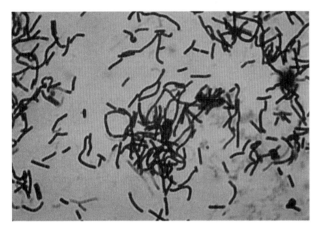

Figure 32 *Erysipelothrix rhusiopathiae* Smear from rough-colony morphotype showing predominance of evenly staining, slightly curved filaments singly, and in short chains

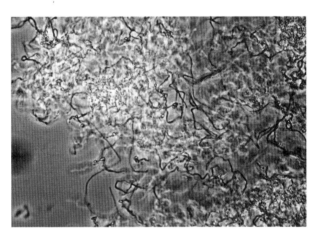

Figure 35 *Erysipelothrix rhusiopathiae* Phase-contrast enhanced wet preparation of broth culture of rough morphotype showing intertwining masses of filaments

Figure 33 *Erysipelothrix rhusiopathiae* Small smooth white colonies with entire edge admixed with tiny grayish rough colonies on 5% sheep blood agar

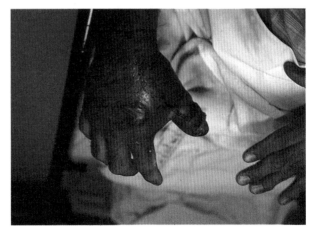

Figure 36 *Erysipelothrix rhusiopathiae* Vesiculobullous violaceous lesion of hand in a fish handler which developed subsequent to fish-bone injury. Note the bruised fingers associated with fish scaling which can serve as a portal of entry for the microorganism

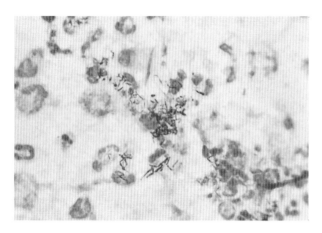

Figure 37 *Erysipelothrix rhusiopathiae* Slightly curved bacilli in smear of aspirate from hand lesion

Figure 38 *Kurthia bessoni* Creamy tan to yellow colonies on chocolate agar after 48-h incubation

Figure 39 *Lactobacillus* Pleomorphic Gram-variable bacilli singly, and in pairs with bulbous swellings after 48 h of growth in liquid media

agar slants or Kligler's iron agar slants, *E. rhusiopathiae* produces a darkening of the butt due to hydrogen sulfide production. This species is catalase-negative.

KURTHIA

Kurthia species are Gram-positive, coryneform, motile rods with parallel sides and a tendency toward coccal formation, especially in aged cultures. *Kurthia* species have been isolated from a variety of sources, such as fresh and putrified meats, waste water, animal feces, and soil. This species occasionally causes endocarditis and bacteremia in patients with underlying disorders who may have indwelling catheters. Infections following traumatic injury in which wounds are contaminated with soil may also occur. Some isolates have been linked to diarrheal episodes.

Morphology

Kurthia species are long rods with rounded ends. In liquid media, filaments are formed which break up into curved chains of coccoid forms. Short rods may predominate in stationary-phase cultures.

Culture characteristics

K. zopfi forms small gray colonies with rhizoid projections emanating from the colony periphery, whereas *K. bessoni* forms yellow-pigmented colonies.

LACTOBACILLUS

Members of this genus are Gram-positive, non-spore-forming, non-motile, pleomorphic bacilli found widely distributed in nature and as inhabitants of the human gastrointestinal tract, mouth and female genital tract. It is a rare human pathogen, occasionally causing septicemia in patients with underlying disorders, e.g. bone marrow transplantation, and endocarditis in patients with pre-existing cardiovascular pathology. Most species are vancomycin-resistant. No virulence factors have been identified.

Morphology

Lactobacilli may present as short bacilli resembling cocci, in chains of varying length, and, in some instances, as circular forms with the ends of individual bacilli in close proximity to each other.

Culture characteristics

Growth is scant on most media. On 5% sheep blood agar, small whitish colonies develop, surrounded by a zone of greenish discoloration (α-hemolysis) of the medium. Cultivation under microaerophilic or anaerobic conditions at 37°C gives optimal growth.

LISTERIA

Listeriae are Gram-positive, non-spore-forming, unencapsulated, short, pleomorphic, motile (25°C) rods which are widely distributed in nature. The optimum temperature for growth is between 30 and

Figure 42   *Lactobacillus* Smear of mucous coating over an intrauterine device (IUD), showing numerous bacillary forms morphologically consistent with 'Doderlein's bacillus' (*Lactobacillus acidophilus*). The IUD was removed in a search for *Actinomyces* species, which is associated with infections in IUD users

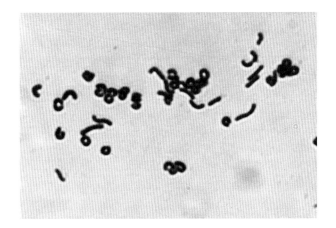

Figure 40   *Lactobacillus* Gram stain of isolate from patient with subacute endocarditis, showing circular forms with ends of individual bacilli almost touching each other

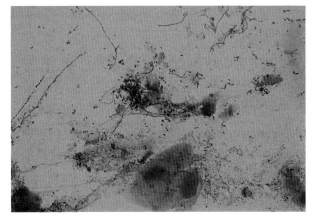

Figure 43   *Lactobacillus* Long, slender, chaining bacilli admixed with polymicrobic flora in direct smear of vaginal exudate

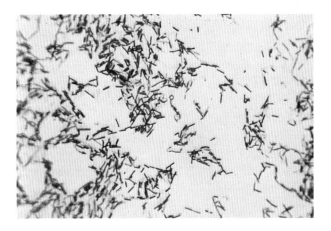

Figure 41   *Lactobacillus* More typical slender bacillary presentation from smear of colonies on 5% sheep blood agar

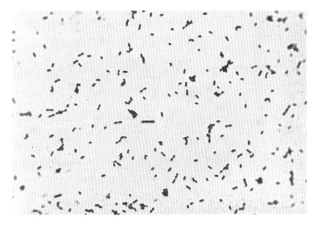

Figure 44   *Listeria monocytogenes* Gram stain of 24-h-old blood agar culture showing predominance of coccal forms and occasional bacillary filament

Figure 45 *Listeria monocytogenes* Gram stain of 48-h-old agar culture showing palisade arrangement resembling *Corynebacterium* ('diphtheroids') species

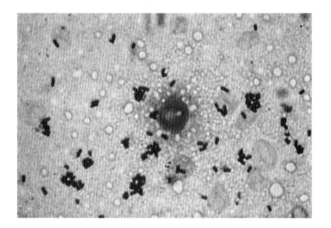

Figure 46 *Listeria monocytogenes* Gram stain of amniotic fluid showing coccoid and small bacillary forms

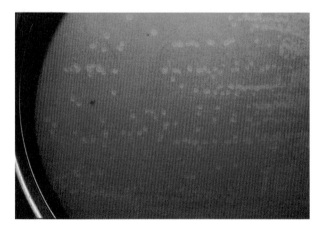

Figure 47 *Listeria monocytogenes* Sheep blood agar culture showing small, smooth colonies encircled by a faint zone of β-hemolysis. Hemolytic activity is better delineated by removing growth with a sterile cotton-tipped applicator and holding a Petri dish up to a light source

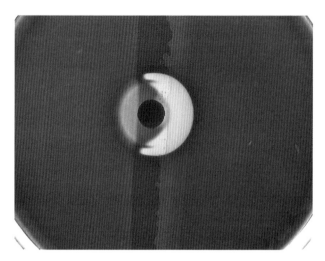

Figure 48 *Listeria monocytogenes* Christie, Atkins, Munch-Petersen (CAMP) test. Both *L. monocytogenes* (left arrowhead) and group B streptococci (right arrowhead) produce the CAMP factor, a diffusible heat-stable protein which enhances the hemolysis of sheep erythrocytes by the β-hemolysin of *Staphylococcus aureus*. β-Hemolysin incorporated into a filter paper disc diffuses onto lawns of Listeria (left) and group B streptococcus (right). Where it interacts with the CAMP protein, a clearer zone of β-hemolysis occurs, manifested as an arrowhead

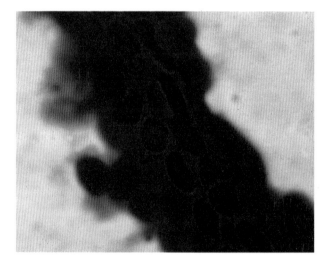

Figure 49 *Listeria monocytogenes* Gram stain of histologic section of placenta showing bacillus transmigrating through placental tissue, facilitating *in utero* infection

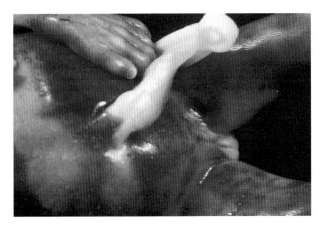

Figure 50 *Listeria monocytogenes* Granulomatosis infantisepticum in still-born fetus showing multiple cutaneous abscesses which were also present on the surface of the liver, spleen, and lungs

Figure 51 *Listeria monocytogenes* Experimental keratoconjunctivitis in rabbit eye subsequent to placing a drop of saline suspension containing *Listeria* onto the conjunctiva. Similar infection will take place in mouse, guinea pig, and human eye. *Listeria* penetrate intact conjunctival epithelium as well as epithelial cells of the urinary bladder and gastrointestinal tract. This phenomenon has been described as 'the epithelial phase of intracellular parasitism'

37°C, but the microorganisms grow well at refrigeration temperature (4°C). The genus contains eight species of which, one, *L. monocytogenes*, is a recognized intracellular human pathogen, causing septicemia and meningitis in immunosuppressed hosts and intrauterine infection in pregnant women, leading to either still birth or congenital early-onset infection. In severe *in utero* infections, disseminated microabscesses (granulomatosis infantisepticum) are found on the liver, spleen, and skin. If acquired during birth from a colonized genital tract, the fetus may develop late-onset infection, usually meningitis, several days after birth. Cutaneous listeriosis may also occur by handling infected animals during parturition. Six serotypes of *L. monocytogenes* have been identified, based on cell wall somatic and flagellar antigens with serovars 1/2a, 4a and 4b, the most common serovars recovered from human infections. *L. monocytogenes* is transmissible to humans mainly through contaminated food and has emerged as a significant cause of food-borne illness. Hematogenous spread from the gastrointestinal tract results in invasion of both phagocytic and non-phagocytic cells. Virulence in *L. monocytogenes* is associated with its hemolysin (listeriolysin), which enables the microorganism to escape from phagosomes and replicate in the phagocyte/macrophage cytoplasm. In this setting, actin is polymerized around the growing bacteria, propelling them through the cytoplasm and thus facilitating cell-to-cell spread. *L. monocytogenes* conjunctivitis and cerebritis have also been described. Non-hemolytic variants of *L. monocytogenes* are avirulent.

Morphology

L. monocytogenes is a pleomorphic bacillus whose morphology can range from short bacillary forms resembling cocci to longer rods and filaments. Coccoid forms predominate in smears of clinical material and in young cultures. Bacilli in palisade formation resembling 'diphtheroids' may also be observed. Older broth cultures may show bacilli in chains resembling lactobacilli. Motility is through peritrichous flagella and is characterized by end-over-end tumbling. It is best observed after growth at 25°C. *Listeria* are catalase-positive and motile, enabling differentiation from group B streptococci (*Streptococcus agalactiae*), which it closely resembles microbiologically and clinically.

Culture characteristics

Listeria monocytogenes grows aerobically and anaerobically on most bacteriologic culture media. On 5% sheep blood agar, colonies are surrounded by a subtle zone of β-hemolysis resembling that produced by group B streptococci. Analogous to group B streptococci, *L. monocytogenes* shows enhancement of the hemolytic activity of the β-hemolysin of *Staphylococcus aureus* (CAMP test).

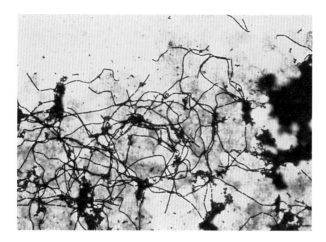

Figure 52 *Oerskovia turbata* Long intertwining filaments in chains in direct smear from blood culture of bacteremic patient

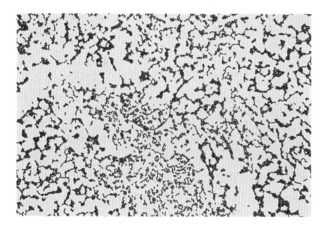

Figure 53 *Oerskovia turbata* Coryneform bacilli with rounded ends, and long filaments in smear of 48-h-old culture on 5% sheep blood agar

OERSKOVIA TURBATA

O. turbata is a Gram-positive, *Nocardia*-like rod that produces branched filaments which fragment into motile rods. *Oerskovia* are soil inhabitants which produce yellow colonies on sheep blood agar. Although they rarely cause human infections, episodes of bacteremia occurred in a 3-year-old boy who had an indwelling Broviac catheter and who had a history of camping outdoors. Bacterial endocarditis in a patient who underwent a homograft replacement of his aortic valve has also been reported.

Morphology

Oerskovia species are nocardioform but may show coryneform morphology, as well as long branched or unbranched filaments in chains after growth in liquid or on agar media.

Culture characteristics

Oerskovia grow on routine bacteriologic media, producing yellow-pigmented colonies after 48-h incubation on heart infusion agar with 5% rabbit blood. Filamentous extensions branch and grow into the agar.

OTHER GENERA

Of the non-acid fast aerobic actinomycetes, members of the genera *Actinomadura*, *Rothia*, and *Streptomyces* have been most frequently associated

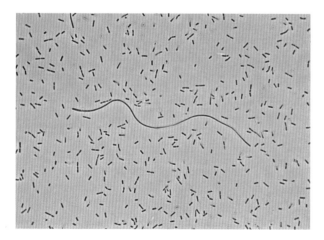

Figure 54 *Oerskovia turbata* Phase-contrast microscopy of wet preparation of agar culture showing 'diphtheroid-like' bacilli and long serpentine filaments

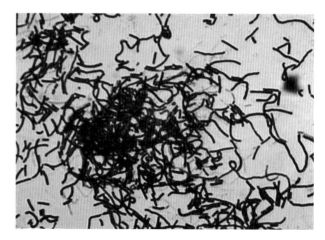

Figure 55 *Rothia dentocariosa* Curved pleomorphic bacilli and filament forms in smear of colony

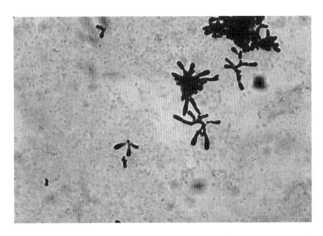

Figure 56 *Rothia dentocariosa* Bacillary forms with distinct clubbing and rudimentary branching

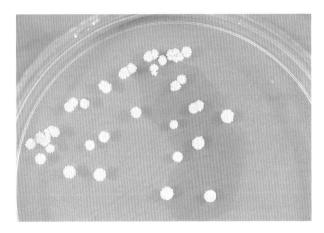

Figure 58 *Rothia dentocariosa* Dry, yellow-orange colonies with serrated borders growing on Sabouraud's dextrose agar

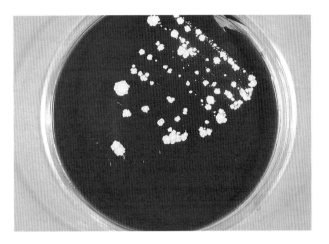

Figure 57 *Rothia dentocariosa* Dry, rough, friable cerebriform colonies with serrated periphery 6–7 days post-incubation at 37oC on 5% sheep blood agar. Note the smaller, pasty colonies which are part of the original inoculum

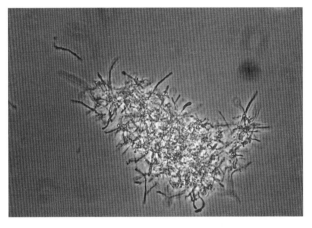

Figure 59 *Rothia dentocariosa* Phase-contrast-enhanced wet preparation of emulsified colony showing densely packed filaments, some with terminal clubbing

with human infections. With the exception of *Rothia dentocariosa*, which colonizes the human oropharynx, *Streptomyces* and *Actinomadura* are mainly soil inhabitants. These species produce filamentous forms with true or rudimentary branching. *Actinomadura madurae* is a frequent cause of actinomycotic mycetomas of the lower extremities following traumatic introduction of the bacterium. *Rothia* is often associated with dental caries and has been reported to cause abscesses and endocarditis.

Morphology

Depending on culture age, *Rothia* are pleomorphic microorganisms ranging from coccoid to filamentous forms, with rudimentary branching and swollen ends. Filaments often fragment into coccoid forms.

Culture characteristics

Rothia grows well on routine bacteriologic media, producing white colonies that are initially rough and crumbly with a cerebriform surface but which may become smooth and easily emulsified after several subcultures. In liquid media, rough colony forms produce a granular growth, whereas smooth colony forms produce uniform turbidity.

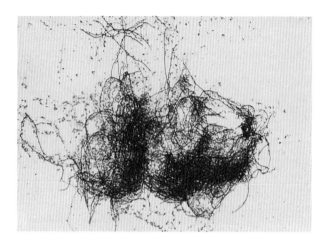

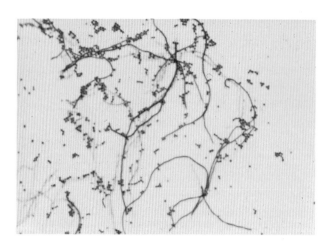

Figure 60   *Streptomyces species*   Tangled mass of filaments from smear of rough, chalky colony growing on Sabouraud's dextrose agar. Colonies have an earthy odor

Figure 61   *Streptomyces species*   Higher magnification showing individual branched filaments with accompanying coccal forms (conidia) disrupted from filaments

2

Gram-positive, aerobic, branching, partially acid-fast species

Bacterial species that are Gram-positive, aerobic, branching and partially acid-fast are members of the order *Actinomycetales* and are commonly found as soil saprophytes. Included among over 40 genera of partially acid-fast bacteria are *Nocardia*, *Rhodococcus*, *Gordona*, and *Tsukamurella*. These genera are enjoined because they contain the glycolipid mycolic acid which accounts for their partial acid-fast staining characteristic. While mycobacteria, e.g. *Mycobacterium tuberculosis*, are also acid-fast, they resist decolorization with 3% hydrochloric acid in alcohol (acid–alcohol-fast), whereas the aerobic actinomycetes do not. Aerobic actinomycetes, however, retain some of the acid-fast stain when decolorized with 1% sulfuric acid. *Corynebacterium* species also contain mycolic acid in their cell wall but do not exhibit either true or partial acid-fast staining because of differences in their mycolic acid chemical structure. Aerobic actinomycetes exhibit delicate branching filamentous forms, which may extend upward (aerial hyphae) from a growing colony. Fragmentation of the filaments into small rods and cocci is common. Human infections in susceptible hosts usually follow inhalation of fragmented filaments or direct subcutaneous inoculation through trauma.

NOCARDIA

The genus *Nocardia* contains at least 11 species, of which *N. asteroides* and *N. brasiliensis* are the most common human isolates. Histologically, human infections are characterized by a marked inflammatory response and granuloma formation. Infections may be localized or disseminated and usually occur in immunocompromised individuals. Pulmonary infection is common, with a propensity for secondary spread of the microorganism to the brain and skin. Primary cutaneous infection can occur in immunocompetent hosts subsequent to traumatic introduction of the bacterium. Rarely, systemic spread may occur from a primary lesion. *Nocardia* are regarded as intracellular pathogens and can survive in phagocytes by inhibiting phagosome–lysosome fusion.

Morphology

Nocardia asteroides forms tangled masses of delicate, Gram-positive, beaded, branched filaments which fragment into bacillary and coccoid forms. Beading

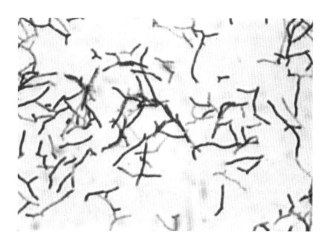

Figure 62 *Nocardia asteroides* Filamentous morphology with rudimentary branching, as seen in smear of agar culture

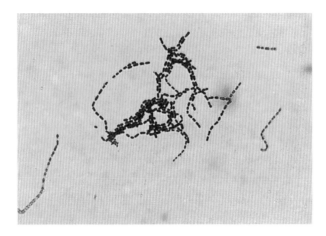

Figure 63 *Nocardia asteroides* Fragmenting filaments into small 'diphtheroid-like' elements

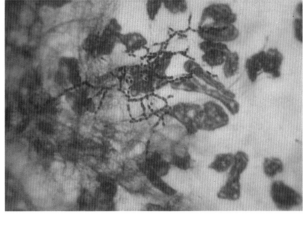

Figure 66 *Nocardia asteroides* Modified acid-fast stain, using 1% sulfuric acid decolorizer, of pulmonary exudate, showing beaded acid-fast (red) filaments

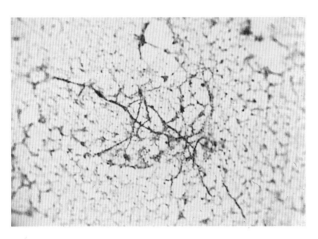

Figure 64 *Nocardia asteroides* Long, branched, beaded, delicate filaments seen in Gram-stained smear of brain abscess

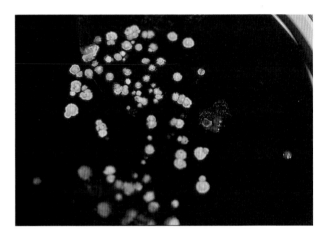

Figure 67 *Nocardia asteroides* Granular white to orange colonies, some with fimbriate (aerial hyphae) borders growing on charcoal-yeast extract (CYE) agar, used for the selective isolation of *Legionella* species. Similar colony morphology may be obtained on other bacteriologic media such as 5% sheep blood agar

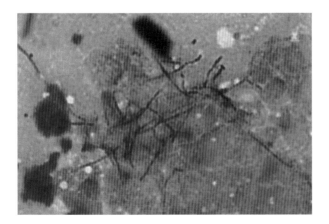

Figure 65 *Nocardia asteroides* Delicate intertwined, beaded filaments in Giemsa stain of bronchoalveolar lavage from immunosuppressed patient with pulmonary nocardiosis

Figure 68 *Nocardia asteroides* Microscopic observation of slide culture on Sabouraud's agar after 48-h incubation, showing delicate, branching filaments coursing over the agar surface from a point inoculum

Figure 69 *Nocardia asteroides* Growth in liquid media as pellicle at broth surface. With time, floccular growth comprising pellicle settles to the bottom of the tube

Figure 70 *Nocardia asteroides Nocardia* possess a lipid-rich cell wall analogous to mycobacteria; they may resist decontamination with 5% sodium hydroxide and grow on Lowenstein–Jensen medium, producing dry, buff to yellow-colored colonies resembling those of *Mycobacterium tuberculosis* with which it may be confused

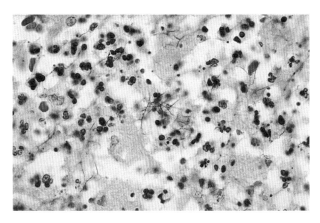

Figure 71 *Nocardia asteroides* Gram stain of histologic section of brain biopsy showing beaded filaments

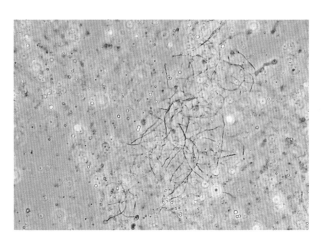

Figure 72 *Nocardia asteroides* Branching, beaded (refractile areas) filaments in phase-contrast-enhanced wet preparation of crushed brain biopsy specimen

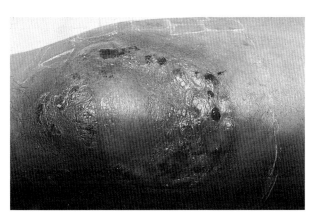

Figure 73 *Nocardia asteroides* Infection of knee graft showing draining, oozing sinus tracts (courtesy of Eric Neibart, MD)

along a bacillary filament bears a striking resemblance to chains of streptococci and may lead to misdiagnosis. Aerial filaments break up into small bead-like spores capable of generating new filaments. *N. asteroides* is weakly acid-fast by the modified Kinyoun stain, using 1% sulfuric acid as the decolorizing agent. Such preparations show both acid-fast (red) and non-acid-fast (blue/green) bacilli and filaments.

Culture characteristics

Nocardia grow slowly (48–72 h) on most routine media at 37°C, imparting a 'wet soil-like' odor to

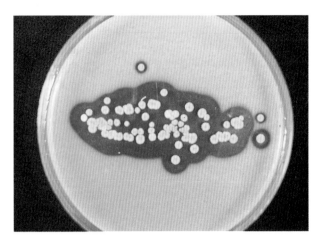

Figure 74 *Nocardia brasiliensis* Hydrolysis of casein is indicated by large zones of clearing around colonies

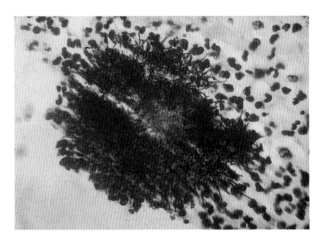

Figure 75 *Nocardia brasiliensis* Mycetoma of foot with multiple draining sinuses which developed secondary to a farming accident in which the foot was pierced and contaminated with soil

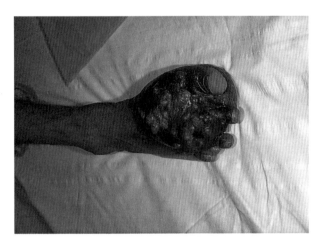

Figure 76 *Nocardia brasiliensis* Actinomycotic-like granule with radiating filaments and pronounced inflammatory response seen in histologic section of foot biopsy

their growth. Colonies are rough and may be chalk white to pinkish and deep orange. Aerial hyphae are present and these allow differentiation of nocardiae from other partially acid-fast species. In liquid media, *Nocardia* produce a granular pellicle at the broth surface, consistent with their aerobic nature. Single floccular granules ('puff balls') may also be observed suspended in the medium. Growth also occurs on Lowenstein–Jensen medium used for the isolation of *Mycobacterium* species. The dry crinkled and slightly buff-colored colonies on this medium mimic those produced by *M. tuberculosis*.

Nocardia brasiliensis

In normal hosts, this *Nocardia* species causes mainly cutaneous infections, usually in the hands or feet, characterized by:

(1) A mycetoma consisting of chronic destructive subcutaneous abscesses with sinus tracts and actinomycotic-like sulfur granules;

(2) A lymphocutaneous infection in which spread is from the point of inoculation up the lymphatics in a 'sporotrichoid' manner;

(3) Superficial cellulitis, and potential systemic spread from a cutaneous lesion.

N. brasiliensis may be differentiated from *N. asteroides* by its hydrolysis of casein, tyrosine, and gelatin. Mycetomas caused by filamentous bacterial species are referred to as actinomycotic mycetomas, in contrast to mycetomas caused by fungi, which are referred to as eumycotic mycetomas. *Nocardia brasiliensis* is most common in Southeast United States.

RHODOCOCCUS EQUI, GORDONA, AND TSUKAMURELLA SPECIES

These environmental species, classified as nocardioform actinomycetes, cause infections among immunosuppressed patients and those with indwelling intravascular catheters. *Rhodococcus equi* is an intracellular pathogen capable of surviving within macrophages by inhibiting phagosome–lysome fusion. It has emerged as a significant cause of pulmonary infection in patients with AIDS. All three Gram-positive species form filaments, some with rudimentary branching, which fragment into 'diphtheroid'-like bacilli and coccobacilli. This microscopic morphology (bacilli to coccoid) is cyclic, depending on incubation time and incubation conditions. Smears of colonies often show a mixture of filaments and coccal forms. *R. equi* produces smooth to mucoid, pink colonies with prolonged incubation. *Gordona* species have caused pulmonary infections, meningitis, catheter-related sepsis, and were incriminated in an outbreak of sternal wound infections following coronary artery bypass surgery. *Gordona* colonies are rough and beige-colored, becoming salmon pink after several days of incubation. *Tsukamurella* species have been associated with catheter-related infections. Colonies resemble those produced by *Gordona* species.

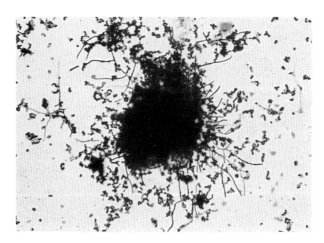

Figure 79   *Gordona species* Modified acid-fast stain of crushed colony showing ramifying acid-fast filaments and short bacillary and coccal forms

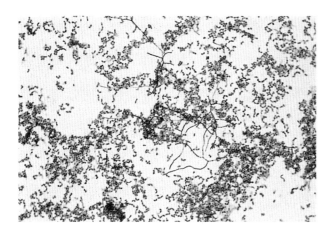

Figure 77   *Gordona species* Gram stain of agar culture showing branched filaments admixed with coccal forms

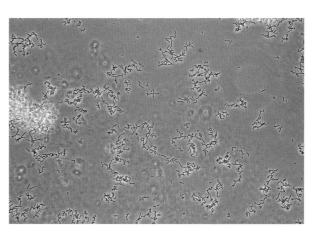

Figure 80   *Gordona species* Phase contrast-enhanced wet preparation of colony fragment showing curved bacilli arranged in Y-forms, V-forms and filaments. Also note the rudimentary branching

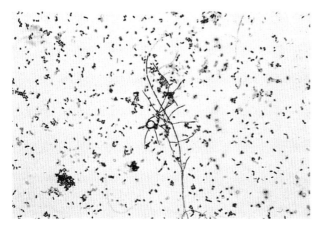

Figure 78   *Gordona species* Modified acid-fast stain showing acid-fast branched filaments and coccal forms

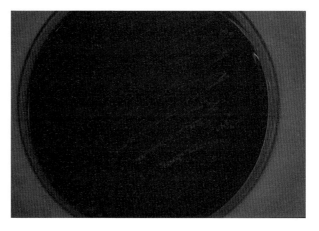

Figure 81   *Rhodococcus equi* Pink-colored colonies on 5% sheep blood agar

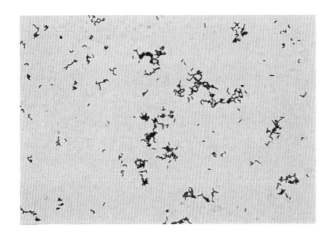 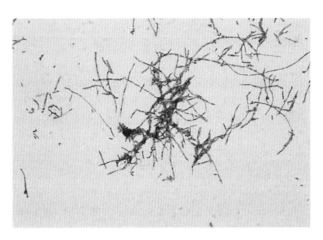

Figure 82 *Rhodococcus equi* Pleomorphic coccobacilli in short filaments, some in palisade formation in smear after 48-h growth on 5% sheep blood agar

Figure 83 *Rhodococcus equi* Beaded partially acid-fast filaments with rudimentary branching in modified (1% sulfuric acid decolorizer) Ziehl–Neelsen stain

3

Acid-fast bacilli: Mycobacterium

All 71 members of the genus are acid–alcohol-fast by virtue of the prodigious lipid content of their cell wall which includes mycolic acid. When stained by the Kinyoun (cold) or Ziehl–Neelsen (hot) method, carbol fuchsin is taken up by mycolic acid and resists decoloration with 3% hydrochloric acid in ethanol. Mycobacteria can also be stained a bright yellow by the auramine-rhodoamine fluorochrome stain. Although mycobacteria are classified as Gram-positive, most species, with the exception of the rapid growers (*M. fortuitum, M. chelonei*), stain poorly by the Gram method, usually appearing as highly refractile, slightly curved, beaded bacilli. Mycobacteria are aerobic, non-spore-forming, and non-motile. Rapid growers may show branching and 'diphtheroid'-like bacilli. Many species produce yellow-pigmented colonies, either upon exposure to light (photochromogens) or in the dark (scotochromogens). *Mycobacterium* species produce a wide spectrum of human and animal infections, both localized and disseminated. For ease of discussion and epidemiologic correlates, mycobacteria can be divided into the *M. tuberculosis* complex (*M. tuberculosis, M. bovis, M. africanum*) and the non-tuberculous mycobacteria, which are not transmissible person to person, and which may colonize healthy individuals. *Mycobacterium leprae*, the causative agent of leprosy, does not grow on bacteriologic media.

MYCOBACTERIUM TUBERCULOSIS COMPLEX

The doubling time for *M. tuberculosis* is 18–24 h. Mycobacteria are intracellular pathogens which can invade and survive in macrophages. Once internalized in a phagolysosome, virulent mycobacteria inhibit acidification of the phagosome and escape into the macrophage cytoplasm, by lysis of the phagolysosome membrane through the action of a hemolysin. Additionally, some lipid components of the mycobacterial cell wall may be toxic to macrophages. *M. tuberculosis* pulmonary infection is acquired primarily by the airborne route from an individual with active pulmonary tuberculosis. Subsequent to the deposition of bacilli in the alveoli of the lungs, macrophage phagocytosis and survival, an inflammatory focus is established which ultimately leads to granuloma formation, characterized by the presence of multinucleated giant cells of the Langhans' type. Now called a tubercle, the focus becomes surrounded by fibroblasts and its center undergoes casseous necrosis.

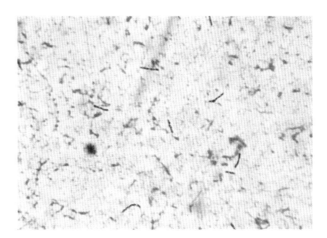

Figure 84 *Mycobacterium tuberculosis* Acid-fast stain of sodium hydroxide-decontaminated sputum specimen from patient with pulmonary tuberculosis, showing typical, slightly curved, beaded bacilli

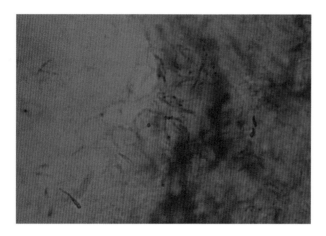

Figure 85 *Mycobacterium tuberculosis* Acid-fast bacilli in direct smear of sputum specimen

Figure 88 *Mycobacterium tuberculosis* Yellow to buff-colored colonies on Lowenstein–Jensen medium

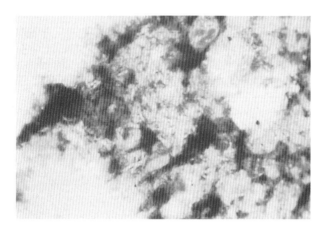

Figure 86 *Mycobacterium tuberculosis* Gram stain of liver abscess. Note carefully the unstained bacilli whose morphology is highlighted by exudate in which they are embedded. This presentation served as the first suggestion of the presence of a *Mycobacterium* species. A separate acid-fast smear of same specimen revealed acid-fast bacilli, and culture grew *M. tuberculosis*

Figure 89 *Mycobacterium tuberculosis* Dry, crinkled, heaped-up, 'cauliflower-like' colonies on Middlebrook 7H10 serum-based medium

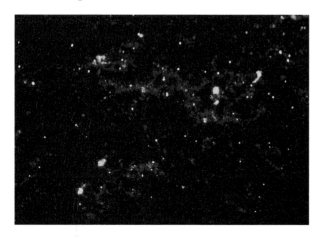

Figure 87 *Mycobacterium tuberculosis* Apple-green to yellow fluorescing bacilli in sputum smear stained with rhodamine–auramine. Low-power scanning of the slide allows for rapid detection of fluorescent bacilli

Figure 90 *Mycobacterium tuberculosis* Granular growth in Middlebrook 7H9 liquid medium. Granules represent individual colonies and are comprised of tightly interwoven bacilli in cord formation

Morphology

The bacilli are either straight or curved to elongated rods, often occurring singly or banded together (cording) because of their lipid-rich hydrophobic surface.

Culture characteristics

M. tuberculosis requires enriched media for growth, as provided by Lowenstein–Jensen (egg-based) and Middlebrook 7H10 and 7H11 (serum-based) media.

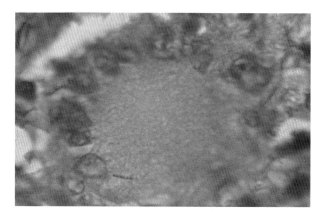

Figure 93 *Mycobacterium tuberculosis* Acid-fast bacillus within multinucleated Langhans' giant cell

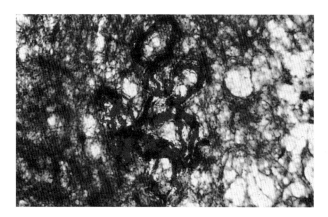

Figure 91 *Mycobacterium tuberculosis* Long interwoven serpentine rows of tightly banded acid-fast bacilli in cord formation. This phenomenon may be observed in smears of liquid media or agar cultures. Cording is not restricted to *M. tuberculosis* and may be observed to a lesser degree with other mycobacterial species, and in direct smears of clinical material

Figure 94 *Mycobacterium tuberculosis* Florid pulmonary tuberculosis with multiple cavities filled with caseous necrotic exudate

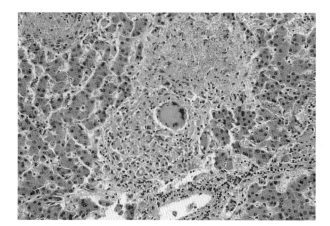

Figure 92 *Mycobacterium tuberculosis* Hematoxylin and eosin stain of histologic section of lung, depicting multinucleated Langhans' giant cell in the center of the granuloma, encircled by a rim of lymphoid cells. Langhans' giant cells have a number of nuclei arranged along the inner periphery of the cell

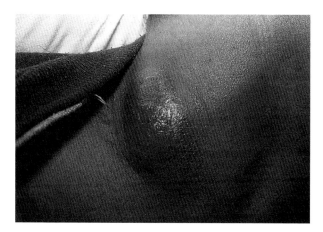

Figure 95 *Mycobacterium tuberculosis* Tuberculous cervical lymphadenitis (scrofula) in patient emigrating from Ecuador. The drainage smear and culture were positive. The patient developed a necrotic lesion at the site of purified protein derivative implantation, indicative of extensive tuberculosis

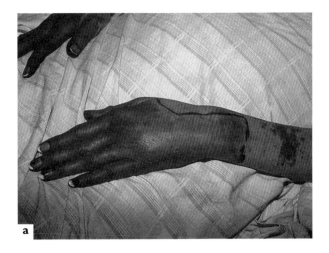

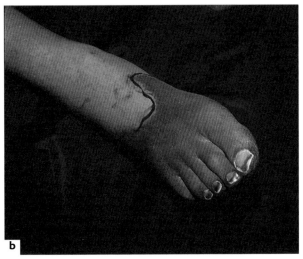

Figure 96 (a) and (b) *Mycobacterium tuberculosis* Disseminated intravascular coagulation in liver transplant recipient with overwhelming tuberculosis (courtesy Burt R. Myers, MD)

On these media, colonies of *M. tuberculosis* develop after 2–3 weeks' incubation at 37°C. Colonies are dry, crinkled, and beige-colored with an irregular surface and contours resembling a cauliflower. This colony topography of *M. tuberculosis* is reflective of its lipid content, which imparts a hydrophobic aspect to colonies. In liquid media, growth occurs in the form of granules, composed of strands (cords) of interwoven bacilli. One commonly used automated liquid media system uses ^{14}C-labelled palmitic acid which is metabolized by mycobacteria releasing radioactively labelled $^{14}CO_2$ which is detected and quantitated, rendering a growth index. An index greater than or equal to 10 is considered positive. *M. tuberculosis* is niacin-positive, a major identifying characteristic.

NON-TUBERCULOUS MYCOBACTERIA (SELECTED SPECIES)

Mycobacterium avium complex

Mycobacterium avium complex (MAC) organisms are widespread in the environment, especially in soil and water. Infection is through inhalation of infected aerosols and by ingestion of contaminated foods. Pulmonary and disseminated MAC infections gained prominence in the setting of patients with AIDS. MAC complex organisms can also cause pulmonary disease and cervical lymphadenitis in normal hosts, particularly in children less than 5 years of age. Colonies of MAC organisms are usually non-

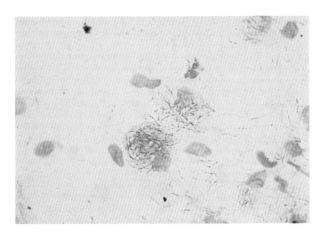

Figure 97 *Mycobacterium tuberculosis* Innumerable intracellular and extracellular acid-fast bacilli in lymph node imprint from liver transplant patient with disseminated intravascular coagulation

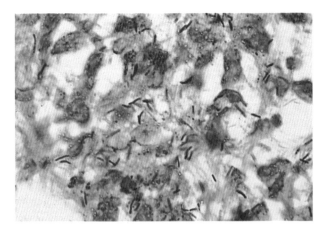

Figure 98 *Mycobacterium avium complex* Numerous slightly curved, beaded bacilli in acid-fast stain of histologic section of skin biopsy

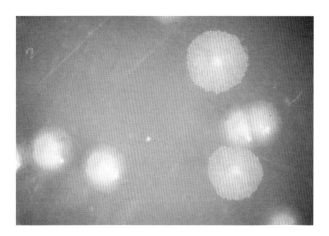

Figure 99   *Mycobacterium avium complex*   Non-pigmented, smooth colonies with spreading fimbriate periphery. Colonies may become yellowish with continued incubation

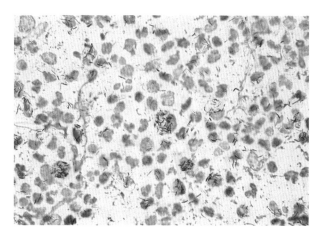

Figure 102   *Mycobacterium avium complex*   Smear of aspirate from a finger lesion, showing numerous acid-fast bacilli singly and intracellular in tissue macrophages

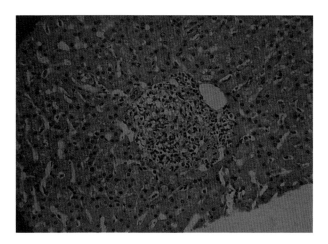

Figure 100   *Mycobacterium avium complex*   Hematoxylin and eosin stain of liver biopsy of AIDS patient with disseminated infection, showing poorly developed granuloma formation

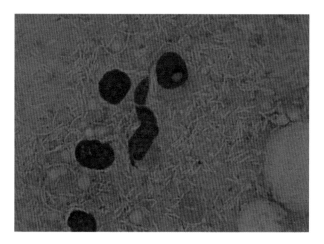

Figure 103   *Mycobacterium avium complex*   Giemsa stain of lymph node imprint, revealing innumerable unstained bacilli

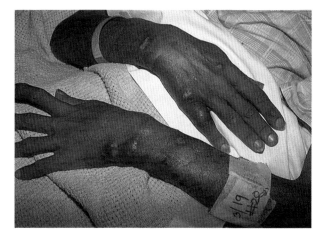

Figure 101   *Mycobacterium avium complex*   Multiple cutaneous abscesses in an AIDS patient with disseminated infection. Note the beginning of lymphatic spread along the small finger

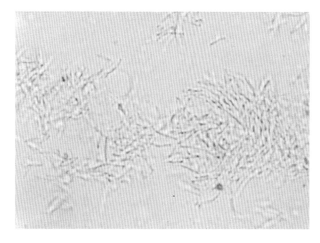

Figure 104   *Mycobacterium avium complex*   Phase-contrast-enhanced wet preparation of crushed lymph node fragment, revealing densely packed, banded bacillary forms

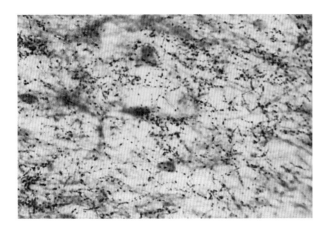

Figure 105 *Mycobacterium avium complex* Innumerable, beaded, acid-fast bacilli in smear of lymph node imprint. Note the absence of tissue cells, indicative of extensive destruction of lymph node architecture

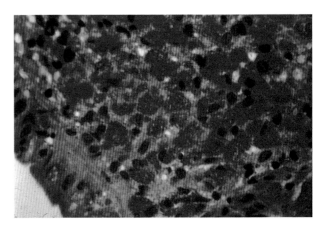

Figure 108 *Mycobacterium avium complex* High-power magnification of colonic biopsy, showing bundles of acid-fast bacilli mimicking those seen in histologic sections of patients with lepromatous leprosy

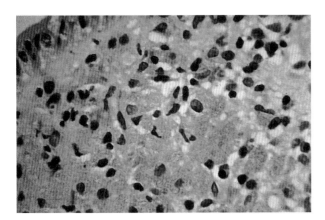

Figure 106 *Mycobacterium avium complex* Hematoxylin and eosin stain of an colonic biopsy of AIDS patient with protracted diarrhea, showing marginal inflammatory response and absence of granuloma formation. Bluish-lavender haze in background is indicative of innumerable poorly staining mycobacteria

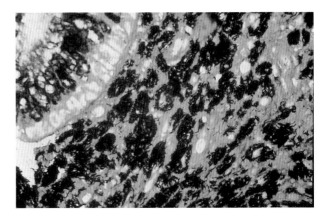

Figure 109 *Mycobacterium avium complex* Methenamine silver stain of colonic biopsy, showing uptake of silver stain (black) by bundles of mycobacteria

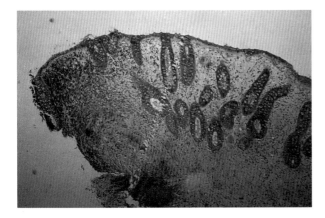

Figure 107 *Mycobacterium avium complex* Low-power view of acid-fast stain of colonic biopsy, showing dense accumulations of acid-fast bacilli

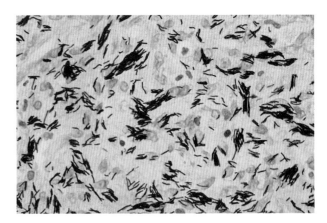

Figure 110 *Mycobacterium avium complex* Bundles of acid-fast bacilli in histologic section of Kaposi's sarcoma in a patient with AIDS. Note the spindle cells in the background, characteristic of histologic presentation of Kaposi's sarcoma

Figure 111 *Mycobacterium kansasii* Dry, rough, colonies which developed yellow pigment 24 h after exposure to light (photoactivation). Prior to light exposure, colonies had the same morphologic texture but were pale buff in color

pigmented, although a rare yellow-pigmented colony form may be encountered. Two colony morphotypes, rough and smooth, are especially evident on serum-based media such as Middlebrook 7H10 and 7H11 agar.

Mycobacterium kansasii

This species is photochromogenic, producing bright yellow colonies upon exposure to light. *M. kansasii* is found in tap water and has caused pulmonary, cervical lymphadenitis (especially in children), cutaneous, and disseminated infections.

Mycobacterium marinum

This species is also photochromogenic. Its natural reservoirs are fresh and salt water, which generally

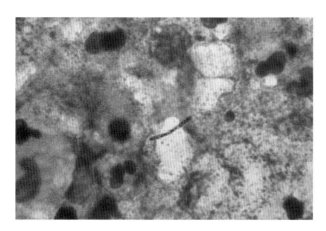

Figure 112 *Mycobacterium marinum* Elongated, beaded acid-fast bacillus in smear of surgical specimen of hand ulcer which developed after fish tank contact

Figure 114 *Mycobacterium marinum* Slender, tapered, beaded, acid-fast bacilli in cord formation seen in smear of colonies

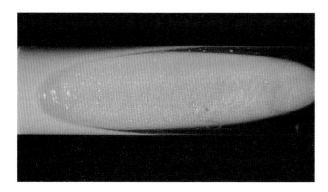

Figure 113 *Mycobacterium marinum* Yellow-pigmented, smooth and rough colonies on Lowenstein–Jensen medium grown at 32°C and exposure to light source, inducing (photoactivation) pigment production

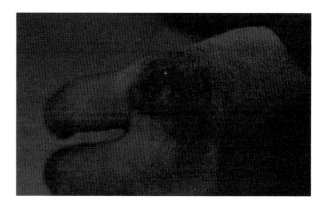

Figure 115 *Mycobacterium marinum* Localized, ulcerated lesion which developed after splinter injury and fish-tank contact

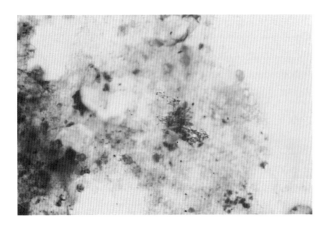

Figure 116 *Mycobacterium marinum* Slender, beaded acid-fast bacilli in smear of fish-tank water in which patient placed hand subsequent to splinter injury

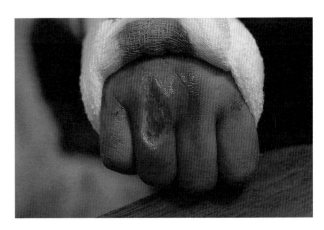

Figure 118 *Mycobacterium marinum* Site of an initial ulcerative lesion in an 8-year-old girl. The lesion underwent multiple debridements and antibiotic treatment for presumed staphylococcal infection. History of exposure to brother's fish tank ('chasing his Guppies') not elicited until microbiologic diagnosis established

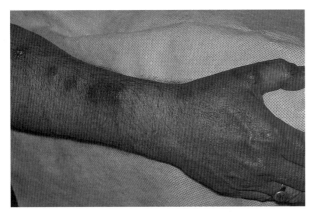

Figure 117 *Mycobacterium marinum* 'Sporotrichoid' progression of abscessed lesions along the lymphatics from primary site of water-related injury to the thumb of a marine biologist

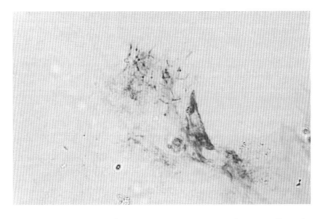

Figure 119 *Mycobacterium marinum* Cluster of acid-fast bacilli in skin scraping of patient's finger

become contaminated by marine life. *M. marinum* produces predominantly deep cutaneous lesions, in association with water contamination, of puncture wounds or abrasions which may occur while swimming, e.g. 'swimming pool granuloma', or handling fish tanks. While most lesions are singular and ulcerative, lymphatic spread in a 'sporotrichoid' manner is not uncommon. Isolation of this species requires incubation of specimens at 30°C.

Mycobacterium scrofulaceum

This species is most commonly linked to cervical lymphadenitis (scrofula) in children. It is found in soil and water and in raw milk and dairy products.

Figure 120 *Mycobacterium scrofulaceum* Beaded, slightly elongated, acid-fast bacilli with modest cording

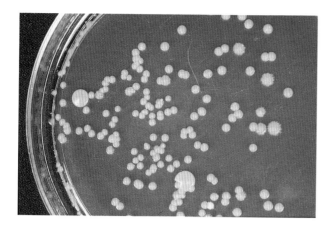

Figure 121 *Mycobacterium scrofulaceum* Yellow-pigmented, smooth and rough colonies with fimbriate projections on Middlebrook agar. Yellow pigmentation developed in both light and dark

This species is characterized by the production of yellow-pigmented colonies in the absence of photoactivation and hence is referred to as a scotochromogen.

Mycobacterium xenopei

This scotochromogenic species is found in tap water and other water sources, as well as among birds, and has been associated with pulmonary and extrapulmonary infections, usually in the setting of individuals with underlying diseases, e.g. malignancies.

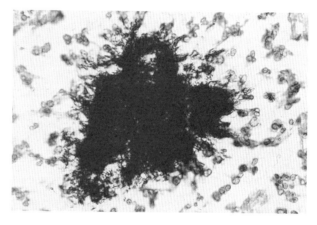

Figure 123 *Mycobacterium xenopei* 'Colony' of densely packed acid-fast bacilli in touch imprint of pulmonary hilar lymph node in patient with diabetes mellitus and chronic myelogenous leukemia who developed disseminated infection

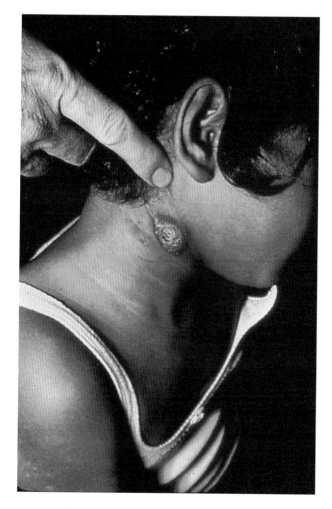

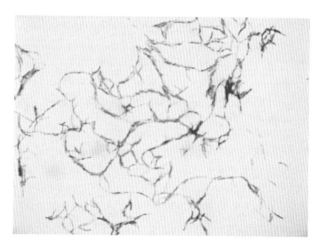

Figure 124 *Mycobacterium fortuitum* Elongated, curved, cording, acid-fast bacilli displaying rudimentary branching in smear of 72-h-old culture growing on 5% sheep blood agar

Figure 122 *Mycobacterium scrofulaceum* Submandibular lymphadenitis (scrofula) with draining sinus tract

39

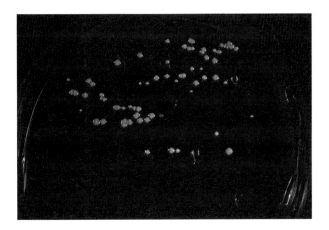

Figure 125 *Mycobacterium fortuitum* Smooth, glistening colonies on 5% sheep blood agar after 72-h incubation. Growth has a 'damp soil'-like odor

Mycobacterium fortuitum, Mycobacterium chelonei, Mycobacterium abscessus

These species are ubiquitous in the environment and are common inhabitants of soil and water. They are characterized by their rapid growth in 48–72 h on most routine bacteriologic media, including that used for the isolation of mycobacteria. Growth is accompanied by a 'damp soil' odor. Colonies are non-pigmented, smooth or rough, and growth also occurs on MacConkey agar without crystal violet. *M. chelonei* grows best at 28–30°C. Because of the reduction in the chain length of their cell wall fatty acid composition, these species not only grow more rapidly, but stain weakly Gram-positive and hence can mimic *Corynebacterium* species (diphtheroids).

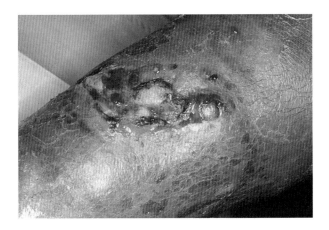

Figure 126 *Mycobacterium fortuitum* Ulcerated lesion of leg with sinus tracts which developed post-cosmetic fat transplant to leg. Similar lesion present in the other leg which underwent the same procedure

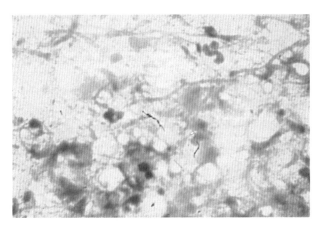

Figure 128 *Mycobacterium chelonei* Acid-fast bacilli admixed with inflammatory response in smear of leg lesion in immunocompetent patient

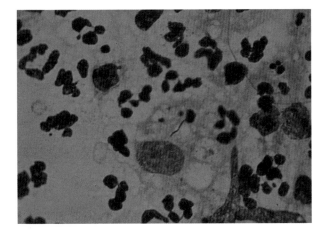

Figure 127 *Mycobacterium fortuitum* Acid-fast bacillus in tissue histiocyte in smear of leg drainage

Figure 129 *Mycobacterium chelonei* Feathery, spreading colonies with rough texture growing on chocolate agar

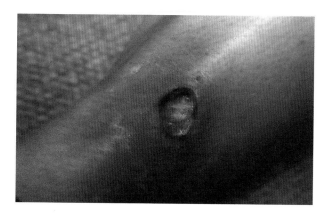

Figure 130 *Mycobacterium chelonei* Abscess of leg with necrotic borders and surrounding erythema in immunocompetent administrative assistant. Shaving of leg, with potential small cuts in skin, may have provided a portal of entry for this environmentally derived microorganism. Culture of in-use shaving cream was negative

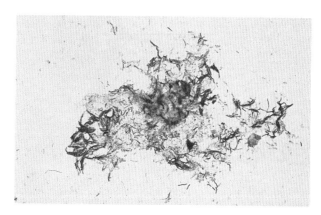

Figure 133 *Mycobacterium chelonei* Gram-positive bacilli in small clusters (banding), some with swollen ends, and beaded filaments, in direct smear of solution from contact lens care system of patient with *M. chelonei* keratitis. Morphologic presentation and staining attributes were highly suggestive for the presence of a rapidly growing mycobacterial species

Figure 131 *Mycobacterium chelonei* Smooth colonies growing on 5% sheep blood agar 48 h after touch inoculation of agar with tissue fragments scraped from leg lesion

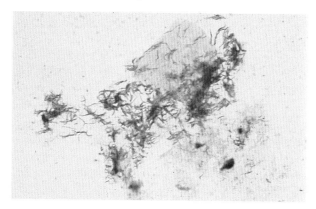

Figure 134 *Mycobacterium chelonei* Innumerable acid-fast bacilli in direct Ziehl–Neelsen-stained smear of solution from contact lens care system of patient with *M. chelonei* keratitis. Culture of solution and eye scraping grew *M. chelonei*

Figure 132 *Mycobacterium chelonei* Dry, rough, crinkled colonies recovered from broviac catheter segment rolled over agar surface. Catheter retrieved from patient with *M. chelonei* cellulitis at insertion site. Patient was 2.5 years old with erythroleukemia

Figure 135 *Mycobacterium chelonei* Growth on MacConkey agar without crystal violet is a characteristic shared with *M. fortuitum* and *M. abscessus*

They can be misidentified or discarded as such. This group of mycobacteria is mainly involved in cutaneous infections in healthy individuals and behaves as an opportunistic pathogen in compromised individuals. *M. fortuitum* and *M. chelonei* have been isolated from corneal ulcers (keratitis). *M. fortuitum* reduces nitrate whereas *M. chelonei* is negative for this characteristic.

Mycobacterium haemophilum

This species requires hemin or ferric ammonium citrate for growth in addition to an incubation temperature of 30°C. This mycobacterium causes mainly granulomatous skin lesions, with subcutaneous abscesses in immunosuppressed individuals. Disseminated infection is rare; it is occasionally associated with cervical lymphadenitis in children. Its natural habitat is unknown.

Mycobacterium leprae

M. leprae is the causative agent of leprosy (Hansen's disease), a chronic disease of the skin, peripheral nerves, and mucus membranes, leading to loss of sensation (anesthesia) and hypopigmentation.

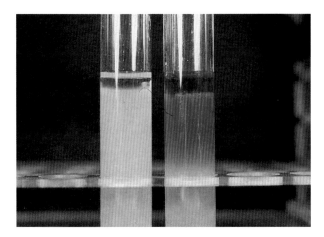

Figure 136 *Mycobacterium chelonei* Positive 48-h arylsulfatase test evidenced by development pink-red color after the addition of sodium bicarbonate to medium containing tripotassium phenothalein sulfate. Arylsulfatase cleaves phenothalein from the tripotassium molecule, which is then detected by the color change upon the addition of sodium bicarbonate. *M. fortuitum* and *M. abscessus* give a similar reaction

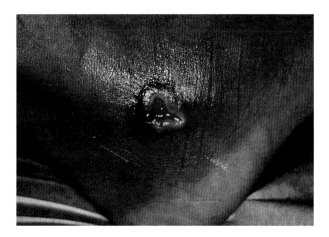

Figure 138 *Mycobacterium haemophilum* Ulcerative lesion on ankle of patient with AIDS draining serosanguinous fluid

Figure 137 *Mycobacterium chelonei* Pellicle formation in liquid media manifested as a thick band of growth with filamentous extensions rising above the liquid surface. With time, the density of the pellicle results in gravitational settling of growth beneath the surface. Pellicle formation also characterizes *Mycobacterium fortuitum* growth in liquid medium

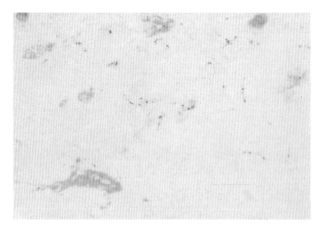

Figure 139 *Mycobacterium haemophilum* Direct smear of draining ankle lesion showing slender, beaded bacilli

Leprosy manifests itself in a tuberculoid or lepromatous form. Tuberculoid leprosy is characterized by the scant presence of acid-fast bacilli in tissues, due to a strong cell-mediated immune response by the host that contains the microorganism. In lepromatous leprosy, the most severe presentation of the disease with numerous skin lesions of the nose and face, the cellular immune response is defective, leading to the unbridled multiplication of acid-fast bacilli in tissue. Leprosy is spread by prolonged person–person contact, through nasal secretions, and skin contact. *M. leprae* does not grow on cell-free media. Diagnosis is achieved by visualization of acid-fast bacilli in biopsy specimens.

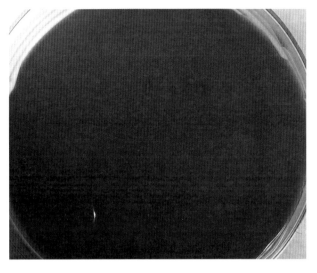

Figure 140 *Mycobacterium haemophilum* Smooth, small colonies developing at inoculation site of drainage onto chocolate agar. In addition to 32°C incubation, this mycobacterial species requires hemin for growth

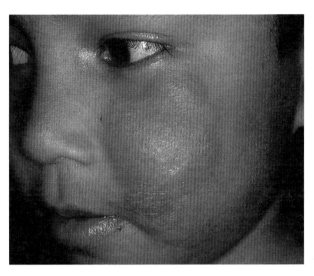

Figure 142 *Mycobacterium leprae* Erythematous facial lesion of leprosy

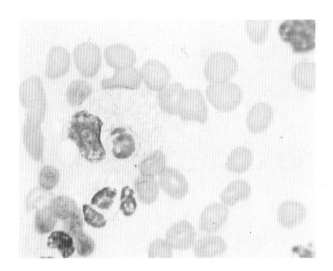

Figure 141 *Mycobacterium haemophilum* Giemsa stain of ankle drainage. Note the faint lavender-stained slender bacilli within phagocytic cell, suggestive of the presence of a mycobacterium

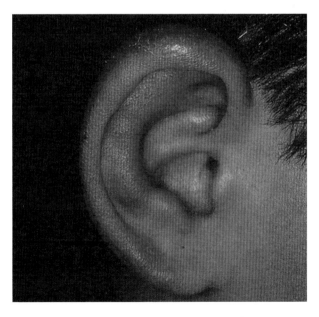

Figure 143 *Mycobacterium leprae* Lesion of the pinna of the ear, which is a common site of involvement because of the preference of microorganisms for a cooler temperature for growth

 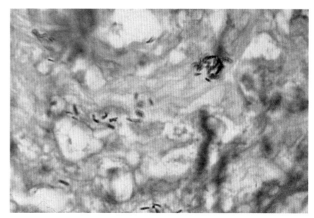

Figure 144 *Mycobacterium leprae* Low-power view of acid-fast stain of histologic section of ear lobe biopsy, revealing innumerable red-staining clusters of bacilli characteristic of lepromatous leprosy

Figure 145 *Mycobacterium leprae* High-power magnification of ear lobe biopsy showing bundles of leprosy bacilli

4

Fermenting, Gram-negative, facultative anaerobic bacilli

ENTEROBACTERIACEAE

Members of this family have a global distribution in soil, water, and vegetation, and comprise part of the gastrointestinal tract flora of humans and animals. Thirty-two genera and more than 130 species have been named. While most species are considered 'normal flora' of the gastrointestinal tract, several bona-fide species are enteric pathogens, e.g. *Salmonella*, *Shigella*, *Yersinia*, and various bio/serogroups of *Escherichia coli*. *Salmonella typhi*, the agent of typhoid fever, and *Yersinia pestis*, the plague bacillus, give localized and systemic disease. Most other members play a major role in hospital-acquired (nosocomial) infections.

Enterobacteriaceae are Gram-negative, motile or non-motile (*Shigella*, *Klebsiella*), non-spore-forming bacilli. All species are oxidase-negative and fermentative, producing acid with or without (anaerogenic) gas from fermentable substrates. *Klebsiella* and *Enterobacter* species are coccobacillary and encapsulated. *Y. pestis* has a pronounced bipolar ('safety pin') staining pattern. Enterobacteriaceae are facultative anaerobes which grow readily on most bacteriologic media, especially selective and differential media containing lactose and bile salts, e.g. MacConkey agar, often producing colonies characterizing the species. Many species, such as the enteric pathogens *Salmonella*, *Shigella*, and *Yersinia*, are non-lactose-fermenting, giving rise to colorless colonies on lactose-containing media. *Proteus mirabilis* and *P. vulgaris* produce non-lactose-fermenting colonies that swarm over the agar surface. Serologic classification is based on somatic O polysaccharides in the outer membrane, flagellar H antigens (motile species), and acid polysaccharides, such as the capsu-lar antigens of *Klebsiella*, the Vi antigen in *S. typhi*, and K antigens in specific *E. coli* strains. K antigens, including the Vi antigen, enhance the virulence of invasive species by inhibiting activation of the alternative complement pathway in non-immune hosts. Depending on the species, Enterobacteriaceae possess a variety of virulence attributes inclusive of adhesins, endotoxin (lipopolysaccharide), hemolysins (*E. coli*), capsules, enterotoxins and cyto-toxins, and serum resistance.

Escherichia coli

This species accounts for the majority of the normal aerobic flora of the gastrointestinal tract. Certain bioserotypes, however, produce gastroenteritis by a variety of mechanisms, e.g. enteroadherence, enteroinvasiveness (similar to *Shigella* species), and

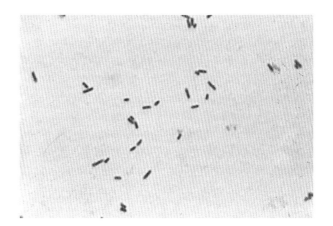

Figure 146 *Escherichia coli* Straight, short, evenly stained bacilli with parallel sides and rounded ends. Coccal and diplobacillary forms also occur. Smear prepared from agar culture

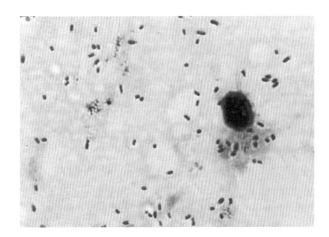

Figure 147 *Escherichia coli* Cerebrospinal fluid smear from newborn with meningitis with short uniformly staining bacilli

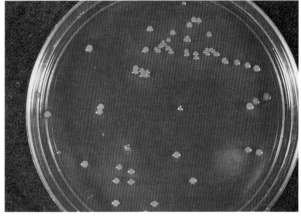

Figure 150 *Escherichia coli* Reddish pink lactose-fermenting colonies on MacConkey agar. Red discoloration surrounding colonies is due to bile salt precipitation from acidic end-products of lactose fermentation

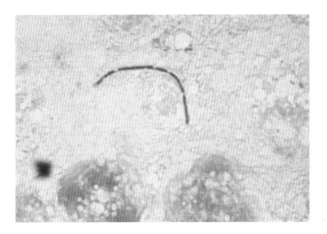

Figure 148 *Escherichia coli* Long bacillary form in smear of thigh abscess aspirate in patient undergoing antibiotic treatment, which accounts for the atypical morphology

Figure 151 *Escherichia coli* Mucoid colony morphotype on MacConkey agar. Such morphotypes may be isolated from patients with chronic urinary tract infections

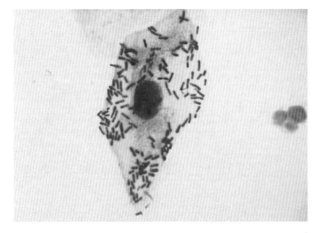

Figure 149 *Escherichia coli* Adherence of bacilli to uroepithelial cell as first step in pathogenesis of urinary tract infection. Specific serotypes frequently associated with urinary tract infections are referred to as uropathic serotypes

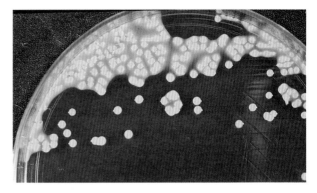

Figure 152 *Escherichia coli* Yellow lactose and sucrose-fermenting colonies on Hektoen–Enteric agar. Acid production from fermentation changes pH indicator brom thymol blue from bluish-green to yellow

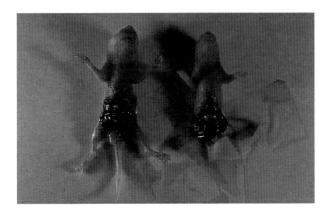

Figure 153 *Escherichia coli* Suckling mouse assay for heat stable (ST) fluid-secreting enterotoxin. Mouse on left orally inoculated with broth culture filtrate of toxigenic strain which resulted in intestinal fluid accumulation, as shown by swelling of intestinal tract. Mouse on left inoculated with toxin-negative isolate. Enterotoxigenic strains are largely responsible for 'traveler's diarrhea'

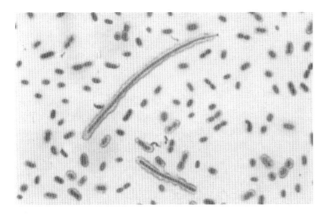

Figure 154 *Klebsiella pneumoniae* Encapsulated bacilli and diplobacilli in smear of 5% sheep blood agar culture. Note that the capsule extends around the entire length of filament forms

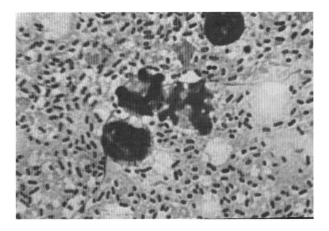

Figure 155 *Klebsiella pneumoniae* Thick, stout, encapsulated diplobacilli in peritoneal exudate of experimentally infected white mouse

enterotoxin production. Heat stabile enterotoxin (which is closely related to the cholera toxin) and heat labile enterotoxin are fluid-secreting (cytotonic), whereas strains such as *E. coli* 0:157 H:7 produce a protein synthesis-inhibiting cytotoxic enterotoxin similar to *Shigella dysenteriae* type 1 enterotoxin. This toxin causes hemorrhagic enterocolitis and the hemolytic uremic syndrome (microangiopathic hemolytic anemia, thrombocytopenia, acute renal failure) with potential central nervous system manifestations. *E. coli* 0:157 H:7 is referred to as enterohemorrhagic *E. coli* (EHEC) and is most often acquired through consumption of undercooked beef products (hamburgers) contaminated with this bacterium. *E. coli* 0:157 H:7 does not ferment sorbitol added to MacConkey agar in lieu of lactose, and therefore can be distinguished from typical sorbitol-fermenting phenotypes. *E. coli* strains with K1 capsular antigen are uniquely associated with neonatal meningitis. *E. coli* is a common cause of urinary tract infections in normal hosts as a consequence of the bacterium's marked affinity for adherence to uroepithelial mucosa. Uropathic *E. coli* strains usually produce β-hemolytic colonies.

Klebsiella

Five species comprise this genus, all of which are encapsulated and non-motile. *K. pneumoniae* (Friedlander's bacllius) causes pneumonia and a variety (wound, urinary tract, septicemia) of nosocomial infections. Pneumonia caused by capsular types 1 and 2 is especially destructive to lung parenchyma with abscess formation. *K. rhinoscleromatis* causes rhinoscleroma, a chronic destructive granulomatous disease of the nasal cavity and pharynx, leading to malformation of the face and neck, and *K. ozaenae* is primarily associated with an atrophic condition of the nasal mucosal membrane, with the secretion of a muco-pus with a fetid odor. *K. oxytoca* is an indole-positive species, causing infections similar to *K. pneumoniae*. Many *K. pneumoniae* isolates produce a plasmid-mediated extended spectrum beta-lactamase which degrades beta-lactam antibiotics. Colonies of klebsiellae are lactose-positive and distinguished by their mucoid, stringing consistency. *Enterobacter* species, a common cause of nosocomial infections, produce similar but less tenacious colonies. *E. sakazaki* and *E. (Pantoea) agglomerans* colonies are yellow-pigmented.

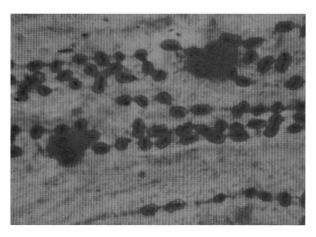

Figure 156 *Klebsiella pneumoniae* Direct smear of sputum from alcoholic patient with pneumonia containing numerous encapsulated diplobacilli. Patient's blood culture was also positive

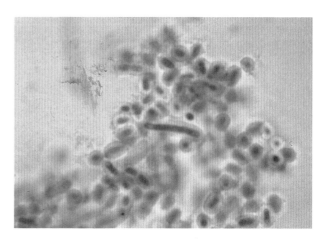

Figure 159 *Klebsiella pneumoniae* Capsular swelling (Quellung reaction) in presence of type-specific anticapsular antiserum, as evidenced by ground-glass appearance of capsule around blue-staining diplobacilli

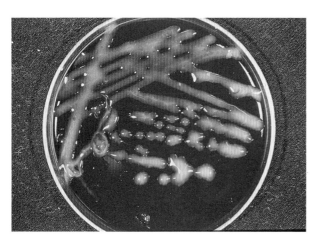

Figure 157 *Klebsiella pneumoniae* Highly mucoid, tenacious, colonies reflective of marked encapsulation. Colonies will 'string' into long threads when an inoculating loop is passed through the colony

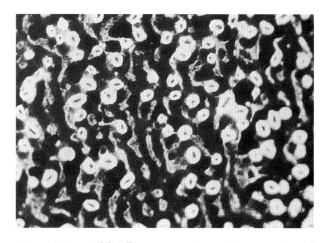

Figure 160 *Klebsiella pneumoniae* Gram stain of Quellung reaction performed by applying type-specific antisera to culture smear and staining after washing. Distinct well-delineated capsules are shown

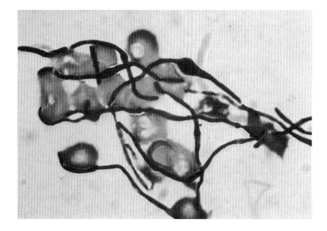

Figure 158 *Klebsiella pneumoniae* Marked pleomorphism consisting of long serpentine filaments with bulbous swellings in smear of blood culture of patient undergoing antibiotic treatment with cell wall-active agent

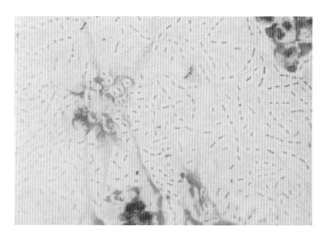

Figure 161 *Klebsiella ozaenae* Encapsulated slender bacilli and diplobacilli in direct smear of purulent sinus aspiration from patient with chronic sinusitis

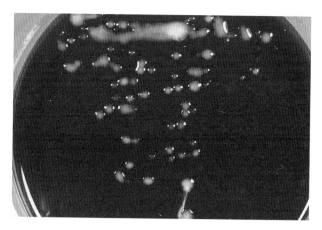

Figure 162 *Klebsiella ozaenae* Mucoid coalescing colonies developing from direct cultivation of sinus exudate onto 5% sheep blood agar

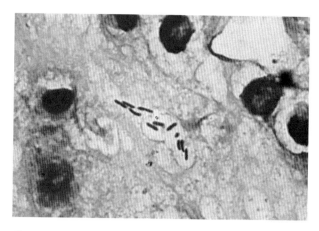

Figure 164 *Pantoea agglomerans* Encapsulated, slightly curved bacilli in direct smear of purulent exudate from brain abscess

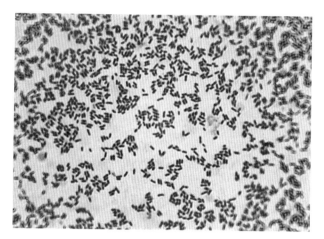

Figure 163 *Pantoea agglomerans* Uniformly staining bacilli and diplobacilli with parallel sides and rounded ends in smear from 5% sheep blood agar culture. Note also the stack (palisade) formation of several bacilli

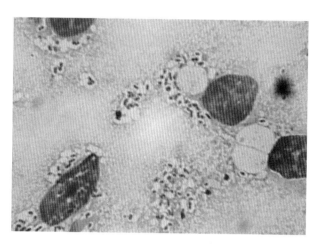

Figure 165 *Pantoea agglomerans* Encapsulated short bacilli and diplobacilli in smear of peritoneal exudate of infected mouse

Pantoea (Enterobacter) agglomerans

Formerly designated *Erwinia herbicola*, this micro-organism has long been recognized as a plant pathogen producing dry necrosis, wilts, and soft rots. Its role as a human pathogen manifested itself during a national outbreak in July 1970 of bacteremic infections associated with contaminated intravenous fluids. This species has since been recovered from the blood of septicemic individuals, usually in the setting of underlying malignancies or immunosuppression. Other associations have included brain abscess, thorn-induced eye and wound injuries, and septic arthritis.

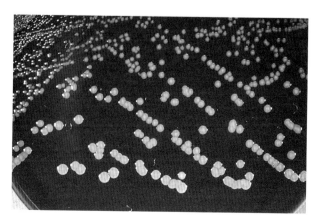

Figure 166 *Pantoea agglomerans* Yellow-pigmented creamy colonies on 5% sheep blood agar after 48-h incubation at 37°C

Figure 167 *Pantoea agglomerans* Yellow-pigmented, smooth, glistening colonies on tripticase soy agar. Pigmentation may be enhanced by incubation at 22°C

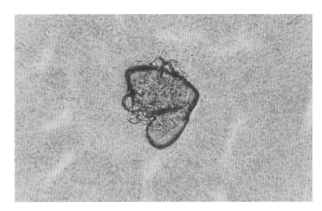

Figure 170 *Pantoea agglomerans* High-power magnification of colony-associated biconvex body which may represent downgrowth of the colony into the agar substrate. Removal of the surface colony growth leaves an agar imprint of the biconvex body

Figure 168 *Pantoea agglomerans* Mucoid, lactose-fermenting, *Klebsiella*-like colony morphotype. This colony form has a rubbery texture and is markedly adherent to the agar surface

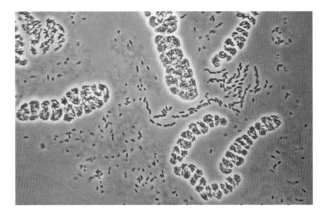

Figure 171 *Pantoea agglomerans* Phase-contrast microscopy of condensate of agar slant culture showing characteristic elongated, spheroidal, aggregated masses of individual bacilli known as symplasmata. This phenomenom, which may also be observed among other plant pathogens, has been attributed to the presence of fibrillar capsular material over the bacterial surface of individual cells which binds them together

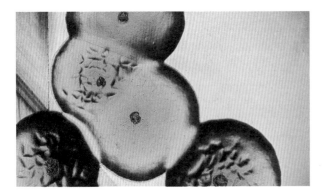

Figure 169 *Pantoea agglomerans* Microscopic view of colonies on clear medium by transmitted light. Each colony shows a centrally situated, biconcave, spindle-shaped body or cellular aggregate, usually developing after 48 h of incubation

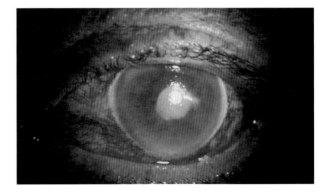

Figure 172 *Pantoea agglomerans* Endophthalmitis secondary to a tree branch injury to eye, with pronounced pus-laden hypoyon

Proteus

Proteus mirabilis and *Proteus vulgaris* comprise the tribe Proteae within the family *Enterobacteriaceae*. These species grow well on most macrobiological media and are distinguished by the production of swarming growth on agar media. Swarming begins at the periphery of colonies or an inoculum and soon encompasses the entire agar surface. Swarming, which is accompanied by a morphological alteration from short bacilli to long filamentous forms and an increase in flagellation, occurs in discontinuous 4–6-h waves of growth and quiescence which form charac-

teristic circular waves. At cessation of swarming, normal bacillary morphology returns. When two different *Proteus* strains are opposed at opposite ends on an agar surface, swarming continues until the swarms meet. The two strains do not coalesae but remain separated by a distinct border. This phenomenon is referred to as the 'Dienes phenomenon' and has been used to track nosocomial episodes of *Proteus* infections. Absence of a line of demarcated spreading is regarded as homology between the strains. *P. mirabilis* and *P. vulgaris* are strong urease producers.

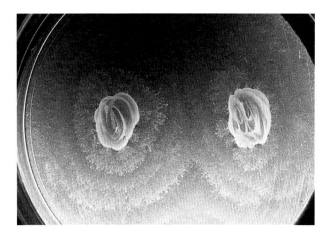

Figure 173 *Proteus mirabilis* Characteristic, merging waves of spreading growth of homologous strains from opposite point inoculations

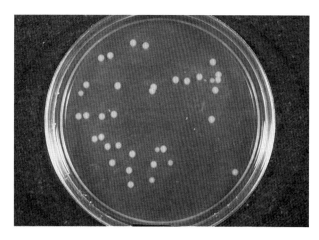

Figure 175 *Proteus mirabilis* Discrete, non-spreading, lactose-negative colonies on MacConkey agar. Swarming is retarded on MacConkey and salt-deficient agar media

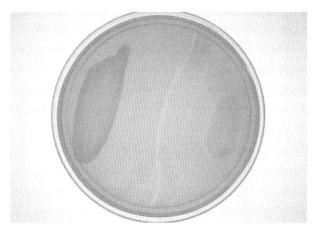

Figure 174 *Proteus mirabilis* Linear zone of swarm demarcation where the swarms of two heterologous strains meet. This phenomenon is known as the Dienes reaction. Often at the zone of interaction, large spheroplast-like bodies (*see* Figure 158) are formed usually by one of the swarming strains. The mechanism underscoring spheroplast formation in this setting is unknown and not related to an antibiotic effect

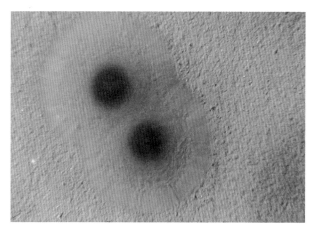

Figure 176 *Proteus mirabilis* High-power magnification of lactose-negative (green) colonies growing on Hektoen–Enteric agar, showing dark central core indicative of hydrogen sulfide production

Shigella species

Shigella are non-motile bacilli and, with the exception of two biotypes of *S. boydii*, do not produce gas from glucose fermentation. Four species comprise the genus: *S. dysenteriae*, *S. boydii*, *S. flexneri*, and *S. sonnei*, the most common isolate in industrialized countries. Occasionally, in the setting of immunosuppression, *Shigella*, especially *S. flexneri*, may invade the bloodstream. *Shigella* species are host-bound to humans and monkeys, and less than 200 organisms are required to establish an infection, thereby facilitating person-to-person transmission via contaminated hands. Other venues include contaminated water and food, especially in countries with poor sanitation. Shigellosis is characterized by watery or bloody diarrhea (dysenteric) with mucoid stools. Shigellae are enteroinvasive and can cause necrosis of the colonic epithelium.

Salmonella

Salmonellae are intracellular pathogens. With the exception of *S. typhi* and *S. paratyphi*, which are host-bound to humans, salmonellae are widely distributed among warm- and cold-blooded animals. Transmission is through food (poultry, eggs, milk) and water contaminated with this species. To date,

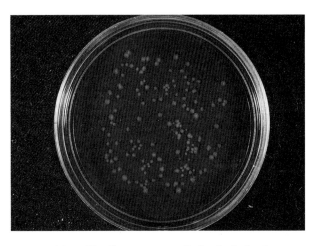

Figure 177 *Shigella species* Pink (colorless), non-lactose-fermenting colonies with serrated contours on MacConkey agar after 24-h incubation at 37°C

Figure 179 *Shigella species* Dysenteric stool characterized by gross blood and mucus. Stool specimen transferred from collection vessel into Petri dish

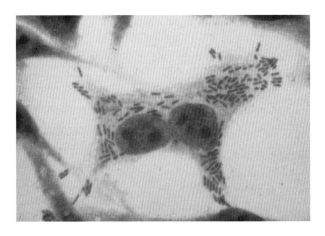

Figure 178 *Shigella species* Invasion of HeLa cells in tissue culture showing distinct alignment of bacilli under the action of actin polymerization, which transposes bacilli through cytoplasm of cell into adjacent cells (courtesy of J. Michael Janda, PhD)

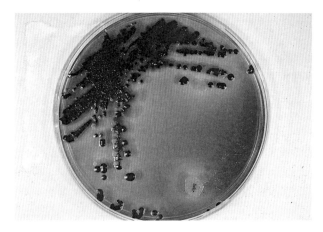

Figure 180 *Salmonella species* Hektoen–Enteric agar culture showing non-lactose-fermenting green colonies with black centers, indicative of hydrogen sulfide production from ferric ammonium sulfate in medium. Colonies of Salmonella typhi (typhoid bacillus) may appear green without dark centers

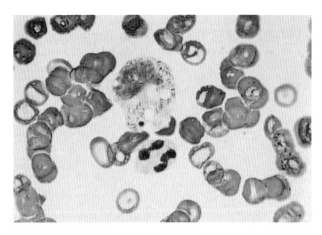

Figure 181 *Salmonella species* Giemsa stain of bone marrow aspirate of child with AIDS showing intracellular bacilli in macrophage. Aspirate was obtained in the search for mycobacterial infection. Culture of the aspirate grew a non-typhoidal *Salmonella* species

Figure 183 *Salmonella species* Mucoid colony morphotype on 5% sheep blood agar. Isolate identified as *Salmonella give*

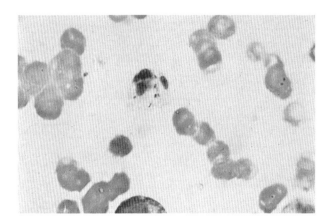

Figure 182 *Salmonella species* Phagocytized Gram-negative bacilli in same bone marrow aspirate, yielding a non-typhoidal species

Figure 184 *Salmonella species* Non-lactose-fermenting mucoid colonies of *Salmonella give* on MacConkey agar

based on serotyping, over 2000 *Salmonella* serotypes have been identified. To ease confusion, the genus *Salmonella* has been divided into seven subgroups with 99% of isolates belonging to subgroup 1, *S. choleraesuis* subspecies *choleraesuis*. This subgroup also includes *S. typhi*, *S. paratyphi*, and *S. typhimurium*. Salmonellosis may be divided into three catergories: gastroenteritis confined to the mucosa and submucosa of the gastrointestinal tract, extraintestinal infection such as bacteremia usually caused by non-typhoidal *Salmonella* serotypes especially in the setting of sickle cell disease, underlying malignancies, chronic liver disease, immunosuppression and the acquired immunodeficiency syndrome

(AIDS), and enteric fever (typhoid) characterized by fever and multisystem involvement. This syndrome is caused mainly, but not solely, by *S. typhi* and *S. paratyphi*, and can be life-threatening. Osteomyelitis and infection of aortic aneurysms can complicate *Salmonella* bacteremia. Gastroenteritis in young children and the elderly may be accompanied by intermittent bacteremia. Salmonellae are lactose-negative and most serovars produce hydrogen sulfide.

Serratia marcescens

This species has been isolated from water, soil, foodstuffs, sewage and animals and has been known to produce disease in horses, rabbits, deer and water

53

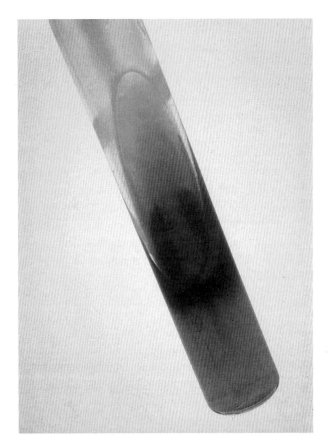

Figure 185 *Salmonella typhi* Characteristic reaction on triple sugar iron agar (TSI) or Kligler's iron agar (KIA) after 24-h incubation showing alkaline (lactose/sucrose negative) slant and acid (glucose fermented) butt without gas, and a rim of blackening (hydrogen sulfide production) at the interface of slant and butt. Most other *Salmonella* serovars blacken the entire butt with hydrogen sulfide (H₂S)

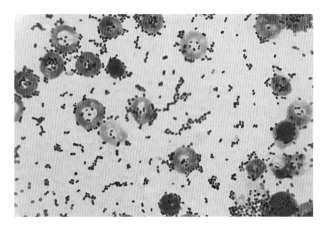

Figure 186 *Serratia marcescens* Small bacilli and coccobacilli in smear of blood culture from leukemic patient housed in an intensive care unit

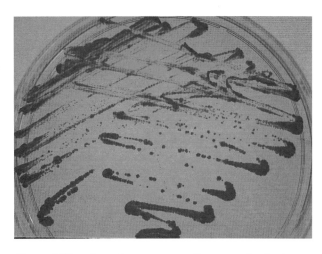

Figure 187 *Serratia marcescens* Orange-red colonies on tripticase soy agar after 24–48-h incubation. On MacConkey agar, red-pigmented colonies may mimick lactose fermentation. This species is lactose-negative

buffaloes. Presently, this microorganism has emerged as a significant nosocomial pathogen, producing a wide range of infectious complications in compromised hosts and bacteremia in intravenous drug abusers. Keratitis subsequent to penetrating eye injuries has also been recorded. While some isolates produce a bright red pigment (prodigiosin) especially at lower incubation temperatures, most isolates are of the non-pigmented variety. This species produces a number of proteases and an extracellular deoxyribonuclease.

Yersinia

The genus *Yersinia* contains 11 species of which three, *Y. enterocolitica*, *Y. pseudotuberculosis*, and *Y. pestis* (plague bacillus), are well-recognized human

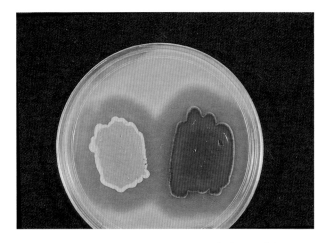

Figure 188 *Serratia marcescens* Hydrolysis of deoxyribonucleic acid by extracellular deoxyribonuclease, as indicated by zones of clearing around spot inoculum of pigmented and non-pigmented isolates

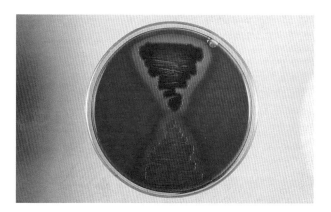

Figure 189 *Serratia marcescens* Zone of β-hemolysis around growth of pigmented and non-pigmented isolates on 5% sheep blood agar

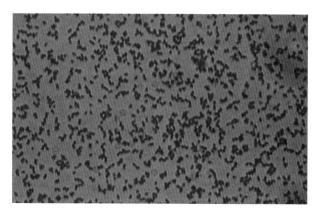

Figure 192 *Yersinia enterocolitica* Small, pleomorphic coccobacilli with rounded ends in smear of 5% sheep blood agar culture. The morphologic presentation of this species can vary with the growth medium. *Y. pseudotuberculosis* gives similar morphology

Figure 190 *Serratia marcescens* Touch imprint onto 5% sheep blood agar of scrub brush retrieved from sink on surgical ward showing innumerable non-pigmented hemolytic colonies. Mosaic pattern reflects inoculation of agar by individual brush bristles

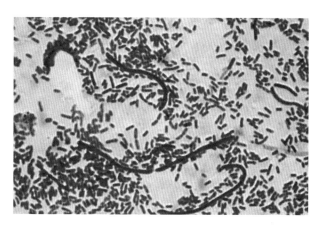

Figure 193 *Yersinia enterocolitica* Pleomorphic bacilli, diplobacilli, and filament forms with parallel sides and rounded ends in smear of growth from a more restrictive medium such as Salmonella and Shigella (SS) agar. Similar morphologic variation may be seen in smears of colonies on Hektoen–Enteric and MacConkey agars

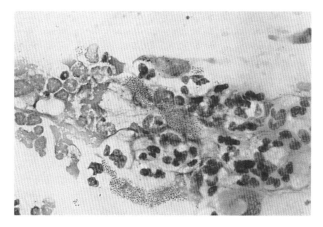

Figure 191 *Serratia marcescens* Innumerable tiny coccobacilli (resembling *Haemophilus influenzae*) in scraping of corneal ulcer in diabetic patient. Culture of scrapings grew red-pigmented isolate

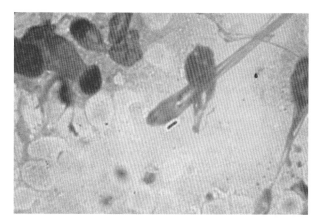

Figure 194 *Yersinia enterocolitica* Encapsulated diplobacilli with rounded ends in smear of peritoneal exudate of infected mouse. *Y. pseudotuberculosis* renders similar morphology and is a natural pathogen of rodents

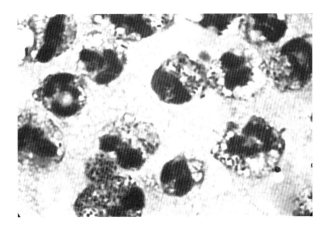

Figure 195 *Yersinia enterocolitica* Intracellular bacilli and coccobacilli in smear of peritoneal exudate of infected mouse. Intracellular survival and growth are a characteristic of *Y. pseudotuberculosis* and *Y. pestis* as well

Figure 197 *Yersinia enterocolitica* Red to burgundy colonies with a transparent periphery on cefsulodin–irgasan–novobiocin (CIN) agar after 48-h incubation at 37°C

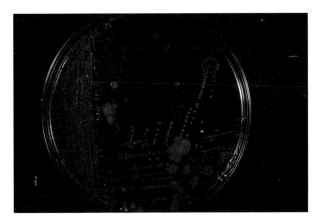

Figure 196 *Yersinia enterocolitica* Small, smooth, non-lactose-fermenting colonies on MacConkey agar after 48-h incubation at 37°C admixed with larger colonies of other enteric flora in direct culture of stool specimen of child with enterocolitis. At 24 h of incubation, colonies are pinpoint on most media and may be overlooked. Children with enterocolitis may have a concomitant bacteremia

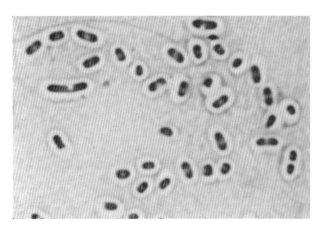

Figure 198 *Yersinia pestis* Encapsulated diplobacilli with pronounced bipolar ('safety pin') staining after growth on serum-enriched medium. Bipolar staining and encapsulation are also apparent in direct smears of infected tissues

pathogens. These species are widely distributed in nature, with certain animal species serving as particular reservoirs: swine for virulent *Y. enterocilitica*, rodents and aviary species for *Y. pseudotuberculosis*, and rodents for *Y. pestis*. *Y. enterocolitica* and *Y. pseudotuberculosis* are acquired mainly from contaminated foods and water, whereas *Y. pestis* is transmitted to humans through flea bites, by direct traumatic inoculation of organisms into the skin or conjunctiva following handling of infected animals, or by aerosols from infected humans or animals (cats) with pneumonic plague. *Y. enterocolitica* and *Y. pseudotuberculosis* are mainly gastrointestinal tract pathogens,

producing enteritis in infants and adolescents, including mesenteric lymphadenitis and terminal ileitis mimicking appendicitis. *Y. pseudotuberculosis* is a natural pathogen of rodents and birds, causing a severe enterocolitis with caseous nodules in Peyer's patches (pseudotubercles). *Y. enterocolitica* and *Y. pseudotuberculosis* are associated with septicemia in patients with iron overload or taking iron chelators such as desferroxamine, and several cases of transfusion-acquired *Y. enterocolitica* septicemia have occurred. *Y. enterocolitica* can grow at 4°C. Plague, a highly fatal infection takes two forms, bubonic and pneumonic, based on the route of acquisition of

Y. pestis. Flea transmission or contact with infected tissue results in bubonic plague and airborne transmission results in pneumonic plague which is highly fatal. Bubonic plague is characterized by the formation of abscesses (buboes) in the lymph nodes closest to a flea bite. Bacteremia may occur with secondary spread to the lungs causing pneumonia. Virulence of yersiniae is associated with both chromosomal and phage-encoded gene products, many of which are expressed at different temperatures depending on the environment encountered, inanimate (food, water), vector (fleas), human or animal hosts. Virulence factors include invasins, antiphagocytic capsules, exoenzymes, serum resistance, outer membrane proteins, and intracellular survival. Yersinae are lactose-negative, motile or non-motile (*Y. pestis*), and capable of growing on most bacteriologic media. Cefsulodin–irgasin–novobiocin (CIN) agar is used for the enhanced recovery of *Y. enterocolitica* which produces dark red colonies on this medium.

VIBRIONACEAE

Members of the family *Vibrionaceae* are all oxidase-positive, motile, glucose-fermenting, aerobic, Gram-negative, straight-to-curved rods, derived mainly from aquatic environments, fresh, brackish or marine. Although none of these microorganisms colonize the human gastrointestinal tract, transient carriage may occur. The genus *Vibrio* contains 12 species of which *V. cholerae*, *V. parahaemolyticus*, and *V. vulnificus* are more commonly occurring gastrointestinal tract pathogens. *V. vulnificus* is additionally a

notorious pathogen, producing septicemia with cutaneous manifestations in compromised hosts, especially those with pre-existing liver disease. The remaining vibrios have less frequently been incriminated in gastroenteritis, wound infections, bacteremia, and ear infections (otitis externa). Although vibrios are halophilic (salt-loving), many species grow on most routine media that contain 0.5% salt concentration. On 5% sheep blood agar and chocolate agar, *Vibrio* colonies are smooth and have a greenish hue. Some species are β-hemolytic,

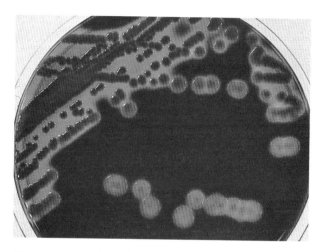

Figure 200 *Vibrio cholerae* Greenish hemolytic colonies of El Tor biotype on 5% sheep blood agar. Isolate recovered from the blood of a cirrhotic patient who consumed crabmeat prior to bacteremic episode. Isolate was shown to be a serogroup 01 strain most often associated with cholera epidemics. Crabmeat may have come from an endemic focus of El Tor biotype along the Gulf Coast of the United States. Rare cases of 01 bacteremia have been reported, usually in the setting of an underlying disorder

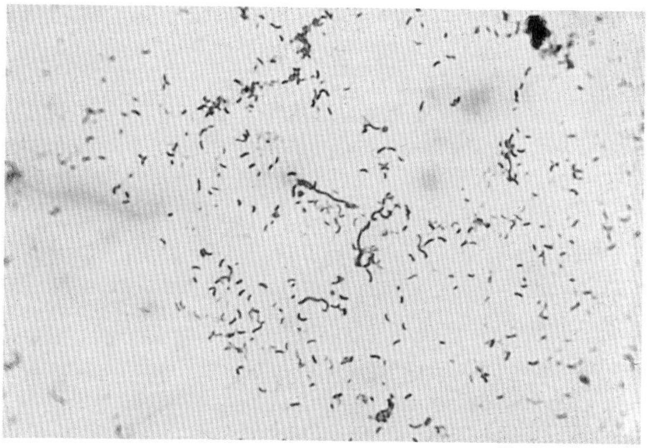

Figure 199 *Vibrio cholerae* Small comma-shaped bacilli and longer curved forms in smear of agar culture

Figure 201 *Vibrio cholerae* Positive string test obtained by emulsifying colonies in 0.5% sodium desoxycholate. Passing an inoculating loop through resulting viscid emulsion results in thread formation

e.g. El Tor biotype of *V. cholerae, V. parahaemolyticus,* and *V. fluvialis.* Thiosulfate citrate bile salts sucrose (TCBS) medium is used for the isolation of vibrios when suspected. This medium allows for differentiation of *Vibrio* species based upon sucrose fermentation (yellow colonies), e.g. *V. cholerae,* form sucrose-negative (green colonies) species, e.g. *V. parahaemolyticus. Vibrio* species are inhibited by the vibriostatic agent 0129, (2,4-diamino-6,7-diisopropylpteridine). *Vibrio* species also render a positive 'string' test after emulsification of colonies in 0.5% sodium desoxycholate. Epidemiologic clues suggesting a *Vibrio* infection include a recent history of raw or undercooked seafood consumption, especially shell fish (oysters), trauma-induced wounds contaminated with fresh or sea water, and foreign travel, particularly to endemic areas.

Vibrio cholerae

Vibrio cholerae serogroup 01, and, more recently, serogroup 0139, cause epidemic cholera, an acute toxin-mediated diarrheal syndrome characterized by profound fluid secretion of one to several liters/hour, up to 20 liters/day, which, if untreated, results in death in hours due to hypovolemic shock. When ingested in contaminated food, mainly raw shellfish, approximately 10^4 cholera vibrios are capable of inducing cholera, an infection of the small bowel where the potent cholera enterotoxin is active. Cholera toxin consists of two major domains, A and B. The B component mediates binding of the complete toxin to intestinal epithelial cell membranes (Gm_1 ganglioside), allowing the A peptide to penetrate the epithelial cells and enzymatically transfer ADP-ribose from nicotinamide adenine dinucleotide (NAD) to guanosine 5' triphosphate (GTP)-binding regulatory protein associated with membrane-bound adenylate cyclase. ADP-ribosylation locks adenylate cyclase in the on-mode, resulting in profuse secretion of Na^+, K^+, Cl^-, bicarbonate (HCO_3^-), and water. The stools of a severely ill cholera patient resemble 'rice water' and contain 10^8 V. *cholerae*/ml being shed into the environment. Direct microscopic examination of a fresh stool will reveal darting, curved bacilli characteristic of *V. cholerae.* Non-serogroup 01 and 0139 *V. cholerae* strains may also give a diarrheal illness by mechanisms distinct from cholera toxin. These strains, however, are non-epidemic. A focus of *V. cholerae* 01 exists along the Gulf coast of the United States.

Vibrio parahaemolyticus

Vibrio parahaemolyticus is a halophilic species associated with gastroenteritis after consumption of raw or inadequately cooked sea food. This species is highly prevalent in coastal waters and may even produce localized skin infections subsequent to trauma. *V. parahaemolyticus* causes either a watery diarrhea or a dysenteric syndrome with blood and mucus in the stools. *V. parahaemolyticus* produces a cytotoxic hemolysin that is closely associated with disease-producing strains. On a high-salt and mannitol-containing blood agar (Wagatsuma agar), virulent *V.*

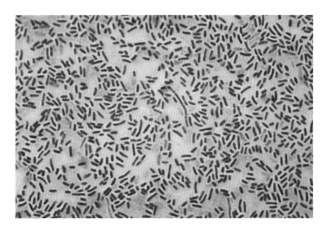

Figure 202 *Vibrio parahaemolyticus* Slightly curved, uniformly staining bacilli with rounded ends in smear from 5% sheep blood agar

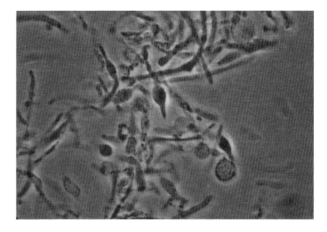

Figure 203 *Vibrio parahaemolyticus* Phase-contrast of wet preparation of agar slant condensate showing marked pleomorphism with filaments containg round (spheroplast) bodies. Pleomorphism with formation of spherical forms may occur spontaneously or induced by cultivation on media with suboptimal sodium chloride concentration

Figure 204 *Vibrio parahaemolyticus* Smooth, glistening, faintly hemolytic colonies with greenish hue characteristic of *Vibrio* species on 5% sheep blood agar

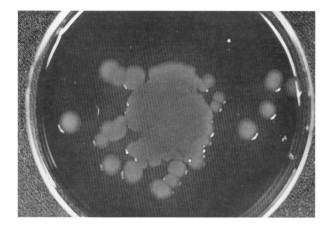

Figure 205 *Vibrio parahaemolyticus* Colonies on human blood agar containing 7% sodium chloride and mannitol (Wagatsuma's agar) showing hemolytic activity termed Kanagawa phenomenom. Virulence of this bacterium is in part associated with hemolysin production

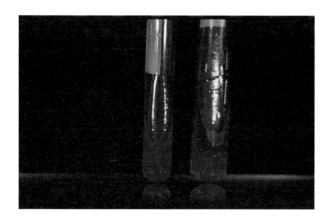

Figure 206 *Vibrio parahaemolyticus* Kliger's iron agar slant (left) showing acid production without gas in butt from glucose fermentation, and alkaline slant indicative of non-fermentation of lactose. Strong lavender reaction on Christensen/urea agar (right) indicative of urease production

parahaemolyticus strains form colonies surrounded by a zone of β-hemolysis, termed Kanagawa phenomenon. Some isolates are urease-positive. An invasive phenotype has also been described.

Vibrio vulnificus

Vibrio vulnificus is a lactose-fermenting halophilic *Vibrio* that accounts for two prominent syndromes, namely wound infections and a life-threatening septicemia in individuals with underlying disorders such as liver pathology and iron overload. Cutaneous manifestations are a prominent component of the septicemic syndrome usually evident in the extremities and trunk. *V. vulnificus* may also cause acute

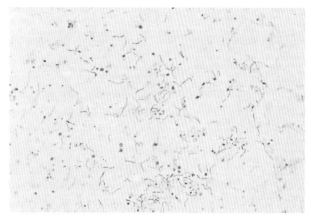

Figure 207 *Vibrio vulnificus* Slender, curved bacilli and many free spherical bodies in smear from 72-h-old agar culture

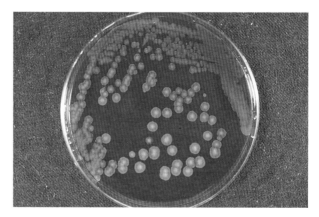

Figure 208 *Vibrio vulnificus* Opaque, silver iridescent non-hemolytic colonies with raised central core on 5% sheep blood agar. Opaque colonies are comprised of cells which have capsular material which is associated with expression of virulence

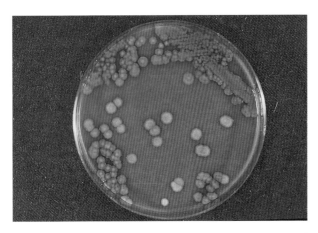

Figure 209 *Vibrio vulnificus* Morphologic variation of colonies on chocolate agar ranging from opaque silver-textured colonies with raised central core to whitish colonies with an irregular contour. Note greenish tint

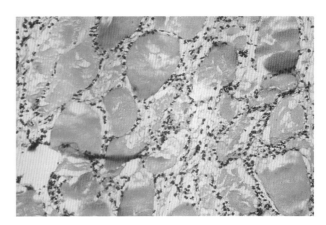

Figure 212 *Vibrio vulnificus* Hematoxylin and eosin stain of histologic section of gangrenous lesion biopsy, showing extensive muscle degradation and innumerable bacilli and coccobacilli coursing between muscle bundles

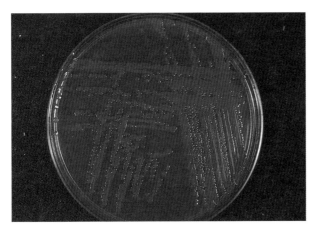

Figure 210 *Vibrio vulnificus* Colorless, non-lactose-fermenting colonies on MacConkey agar after 24-h incubation. Rapidity of lactose fermentation (deep pink to red colonies) may vary between strains

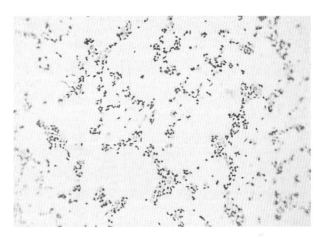

Figure 213 *Vibrio alginolyticus* Short, slightly curved, irregularly staining vacuolated bacilli and coccobacilli in smear of 18-hour-old culture on 5% sheep blood agar

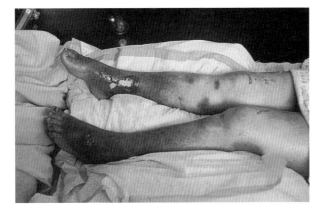

Figure 211 *Vibrio vulnificus* Extensive, bullous and gangrenous lesions with peripheral edema in alcoholic patient who developed bacteremia after consuming raw oysters (courtesy Yin-Ching Chuang, MD)

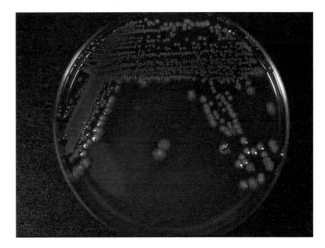

Figure 214 *Vibrio alginolyticus* Smooth, non-hemolytic colonies with raised central core from subculture of positive blood culture of patient with osteogenic sarcoma

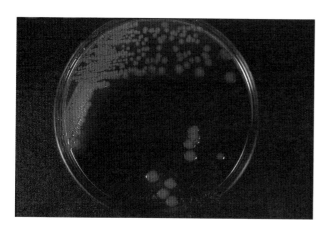

Figure 215 *Vibrio alginolyticus* Yellow–orange colonies indicative of sucrose fermentation on thiosulfate citrate bile salts sucrose (TCBS) agar

Figure 216 *Vibrio fluvialis* Silver iridescent β-hemolytic colonies with raised central core

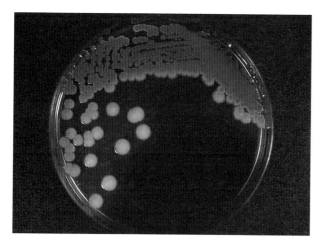

Figure 217 *Vibrio fluvialis* Sucrose-fermenting yellowish colonies on thiosulfate citrate bile salts sucrose (TCBS) medium

diarrhea with, or without septicemia, in individuals on antacid therapy or who have undergone gastrectomies. Because of its lactose-fermenting capability, colonies of *V. vulnificus* on enteric media such as MacConkey agar may be confused with those of *E. coli* and thus potentially overlooked. Other infections associated with this species include corneal ulcer, acute epiglottitis, and endometritis.

AEROMONADACEAE

This family includes *Aeromonas* species and *Plesiomonas shigelloides*, both of which are widely distributed in fresh water, soil, and marine environments. These species are closely related to vibrios in being Gram-negative, oxidase-positive, glucose-fermenting, motile bacilli. The genus *Aeromonas* contains 13 species based on hybridization groups of which *A. hydrophila*, *A. veroni biovar sobria*, and *A. caviae* are the most frequently isolated from human infections. *Pl. shigelloides* is a distinct species with many shared characteristics with aeromonads.

Historically, aeromonads have been recognized as pathogens of warm- and cold-blooded animals (fish, reptiles, mammals, humans). Mounting clinical and experimental evidence supports the role of *Aeromonas* species as enteropathogens. *A. hydrophila* and *A. veroni biovar sobria* produce a cytoxic hemolysin and a cytotonic (fluid-secreting) enterotoxin and adhesins all suspected to play a role in enteropathogenicity. *A. caviae*, while non-hemolytic, also produces a cytotonic enterotoxin. *A. hydrophila*

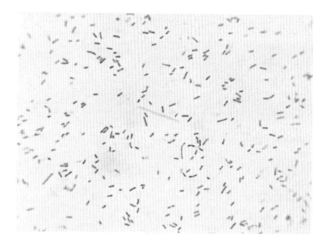

Figure 218 *Aeromonas hydrophila* Short-to-medium length bacilli with parallel sides and rounded ends in smear of growth from 5% sheep blood agar

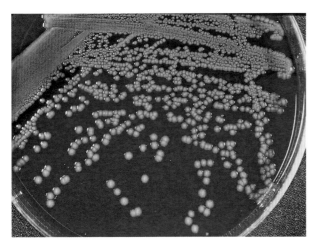

Figure 219 *Aeromonas hydrophila* Smooth, buff-colored β-hemolytic colonies on 5% sheep blood agar. Hemolysis is better visualized by displacement of the colony with a sterile cotton-tipped swab

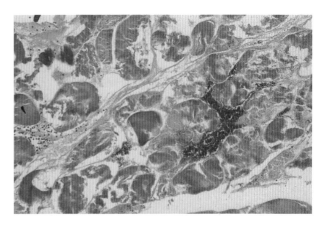

Figure 222 *Aeromonas hydrophila* Hematoxylin and eosin stain of section of rabbit leg biopsy showing muscle degradation, extravasated red blood cells, and definitive lavender-staining band of bacterial cells transecting muscle bundles

Figure 220 *Aeromonas hydrophila* Direct plating of egg salad incriminated in diarrheal illness to 5% sheep blood agar showing innumerable β-hemolytic colonies. The author who conducted the investigation also developed diarrhea after consuming egg salad

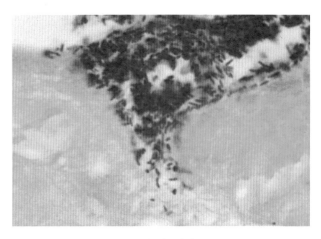

Figure 223 *Aeromonas hydrophila* Gram stain of histologic section of necrotic rabbit muscle revealing dense accumulation of bacilli between muscle bundles, resulting in local anoxia, and synthesis of proteases (collagenase, elastase), all contributing to muscle necrosis

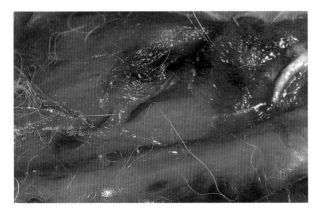

Figure 221 *Aeromonas hydrophila* Hemorrhagic necrotic lesions of leg of bacteremic rabbit. Lower body temperature in extremity favors growth of this species

strains have been linked to a cholera-like diarrhea associated with a serogically cross-reactive cholera toxin-like enterotoxin. Dysenteric-like diarrhea has also been associated with *A. hydrophila*. *Aeromonas* species produce a wide range of exoenzymes, including proteases, such as elastase, thought to play a major role in cutaneous infections, septicemia, and pulmonary disease. Bacteremic aeromonads are serum-resistant. Wound infections in normal and compromised patients are fairly common in association with fresh water trauma and can range from cellulitis to ecthyma gangrenosum. Septicemia may complicate wound infections. *Aeromonas* and *Plesiomonas* grow readily on most culture media,

Figure 224 *Aeromonas hydrophila* Fluid accumulation (bottom) in intestinal tract of suckling mouse fed culture filtrate of enterotoxigenic strain. Note the absence of distended intestine in control mouse fed filtrate from non-toxigenic strain. *A. caviae* also produces a fluid-secreting enterotoxin

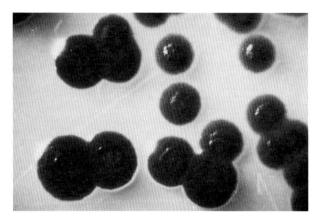

Figure 225 *Chromobacterium violaceum* Round, smooth, lavender colonies with an opaque center and entire edge after 3 days' incubation at 37°C. Colonies may show β-hemolysis on 5% sheep blood agar

producing large, round, glistening colonies with entire edges. On 5% sheep blood agar, *A. hydrophila* and *A. veroni* produce buff-colored β-hemolytic colonies.

Chromobacterium violaceum

Chromobacterium violaceum is a normal inhabitant of soil and water which occasionally produces infections in various animals and humans in southeast Asia and the southeastern United States, especially Florida and Louisiana. The bacterium gains access primarily through penetrating injuries to the skin. Another portal of entry is thought to be through ingestion of, or exposure to, contaminated water. Infectious complications include skin ulcers, septicemia, and abscesses of the liver and lungs, among others. Infections can be severe and fatal. Several cases have been reported in patients with chronic granulomatous disease in childhood. *C. violaceum* ferments glucose.

Morphology

C. violaceum is a slightly curved bacillus with rounded ends arranged singly, and in pairs and which

Figure 226 *Chromobacterium violaceum* Violet collar remaining at surface of turbid broth culture after pellicle has settled to the bottom of the tube

may show bipolar staining. The bacterium is motile by polar and lateral flagella.

Culture characteristics

C. violaceum grows well on most bacteriologic media, producing an ethanol-soluble violet pigment, violaceum, at 37°C. In liquid media, a lavender pellicle is produced.

5

Non-fermenting, Gram-negative aerobic bacilli

Species that fall into this category are Gram-negative bacilli or coccobacilli that can be divided into those that oxidatively break down carbohydrates and those that are non-oxidizers. The species comprising these groups are ubiquitous in nature and are found in water, soil, soil plants, and decaying vegetation. Many of the species contained herein are major nosocomial pathogens. Among the latter, *Pseudomonas aeruginosa*, *Acinetobacter* species, *Stenotrophomonas maltophilia*, and *Burkholderia cepacia* are frequent offenders. Favoring nosocomial spread is the ability of these microorganisms to survive under a variety of harsh environmental conditions, including those in humidifiers and in disinfectant solutions. Additionally, these species have a propensity to colonize the human skin and gastrointestinal tract after prolonged hospitalization, especially in intensive care units, and under antibiotic selective pressure.

PSEUDOMONAS AERUGINOSA

This species is among the most tissue-destructive in patients with underlying disorders, especially immunosuppression, leukemia, neutropenia, and in those who have undergone surgery. *P. aeruginosa* is both invasive and toxigenic in addition to expressing numerous proteolytic exoenzymes, including elastase which disrupts the elastin layer of blood vessels, leading to hemorrhage and spillage of organisms (perivascular cuffing) into the surrounding tissue. Exotoxin A is a major virulence factor which acts on a molecular basis similar to the *Corynebacterium diphtheriae* toxin, inhibiting protein synthesis. After activation and entry of the toxin into susceptible cells, fragment A catalyzes the transfer of ADP-ribose

from nicotinamide adenine dinucleotide (NAD) to elongation factor 2 (EF 2), inhibiting the assembly of polypeptides on the ribosome. In patients with cystic fibrosis, *P. aeruginosa* colonizes the lung in microcolonies surrounded by copious amounts of capsular exopolysaccharide alginic acid. This feature is carried over to *in vitro* growth in the form of highly mucoid colonies almost pathognomonic for cystic fibrosis. Clinically, *P. aeruginosa* can produce a hemorrhagic pneumonia with abscess formation in patients with underlying malignancies. Bacteremia is common in this setting and often accompanied by necrotic cutaneous lesions (ecthyma gangrenosum). Urinary tract infections secondary to catheterization tend to

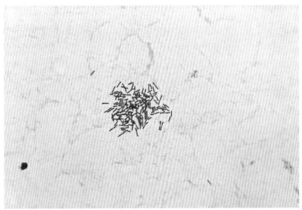

Figure 227 *Pseudomonas aeruginosa* Slender bacilli with parallel sides and rounded ends in smear of cerebrospinal fluid. Direct observation of wet preparation showed active, darting motility, characteristic of a polar flagellated bacterium

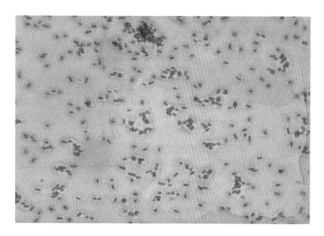

Figure 228 *Pseudomonas aeruginosa* Encapsulated, slender bacilli singly and in pairs in smear of mucoid colony. Encapsulated variants often appear more coccobacillary in smears

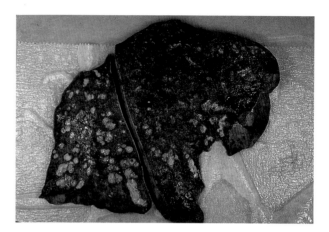

Figure 231 *Pseudomonas aeruginosa* Lung of patient who expired with cystic fibrosis with ectactic small airways that extend to the lung periphery. Air spaces are dilated and filled with mucus and inflammatory response

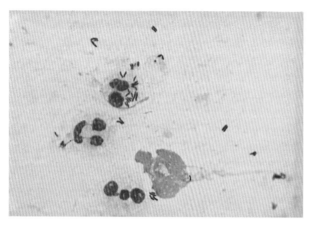

Figure 229 *Pseudomonas aeruginosa* Encapsulated, slender bacilli in smear of sputum of neutropenic patient with upper respiratory tract infection

Figure 232 *Pseudomonas aeruginosa* Microcolonies of several encapsulated bacilli enmeshed in red-staining capsular exopolysaccharide (alginic acid slime layer). This morphologic finding is common in lung secretions of patients with cystic fibrosis and serves to enhance lung colonization by repelling phagocytosis and antibody and antibiotic access to the bacterial cell. Alginic acid also provides a biofilm enabling *P. aeruginosa* to adhere to catheters and invade the human body

Figure 230 *Pseudomonas aeruginosa* Soft agar culture showing blue–green (pyocyanin) pigment and radial migration of motile bacilli from point inoculation. Motility is regarded as a virulence factor, accounting for bacteremic spread in burn patients

become chronic. Burn patients are particularly prone to *P. aeruginosa* superinfection of charred tissue, which often leads to bacteremia. *P. aeruginosa* can also produce a severe keratitis in association with contact lens wear or penetrating eye trauma. *P. aeruginosa* causes a superficial external otitis (swimmer's ear) in normal individuals and a notoriously destructive invasive ear infection (malignant external otitis) with bone destruction in diabetic patients. *Pseudomonas* folliculitis after exposure to water ('splash rash') in whirlpool baths, jacuzzis, and hot tubs is a potential risk of the recreational or therapeutic use of water. Other infectious complications include osteomyelitis subsequent to puncture wounds, meningitis, and endocarditis, predominantly in intravenous drug abusers and in those with prosthetic heart valves.

Morphology

Pseudomonas aeruginosa is a slender, highly motile, non-sporulating, Gram-negative rod with rounded ends. In some preparations, bipolar staining may be evident. Cells occur singly or in small clusters. Motility is rapid through a single polar flagellum. Encapsulation is seen with many strains associated with chronic lung disease such as cystic fibrosis. *P. aeruginosa* is strongly oxidase-positive.

Culture characteristics

P. aeruginosa grows best aerobically on most bacteriologic media, often producing a blue–green diffusible pigment (pyocyanin) and a 'grape-like' odor. *P. aeruginosa* produces several colony morphotypes, ranging from flat colonies with a silvery iridescence with 'patches' spotting the colony surface, to colonies that are smooth and entire or have a spreading character with serrated borders. On 5% sheep blood agar, colonies are β-hemolytic. By tilting the agar medium, a metallic sheen is also visible. Some strains elaborate a brown pigment (pyoverdin). Mucoid *Klebsiella*-like colonies are often recovered from culture of respiratory specimens of patients with cystic fibrosis. In liquid media, a pellicle often forms and, with shaking of the tube, a blue–green pigment will become evident. Blue–green pigmentation may also be observed in clinical specimens, e.g. cerebrospinal fluid, in patients with severe *P. aeruginosa* infections.

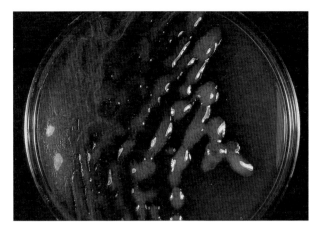

Figure 233 *Pseudomonas aeruginosa* Characteristic blue–green-pigmented mucoid colony morphotype isolated from sputum of patient with cystic fibrosis

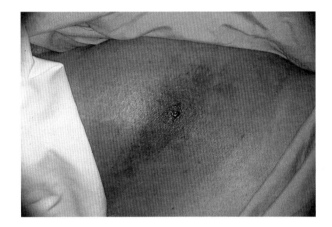

Figure 234 *Pseudomonas aeruginosa* Ecthyma gangrenosum lesion of thigh of neutropenic patient with bacteremia. Necrotic eschar is due to tissue-destructive extracellular exoenzymes, especially proteases and α-toxin

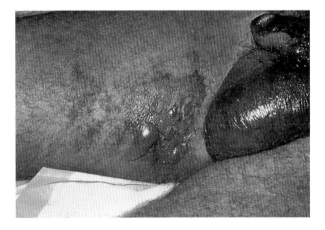

Figure 235 *Pseudomonas aeruginosa* Bullous ecthyma gangrenosum lesion of inner thigh and concomitant gangrenous lesion of penis and scrotum

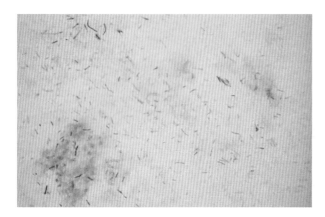

Figure 236 *Pseudomonas aeruginosa* Smear of aspirate of ecthyma lesion showing numerous slender bacilli in absence of inflammatory response. Wet preparation of aspirate showed actively motile bacilli

Figure 237 *Pseudomonas aeruginosa* Slightly spreading blue–green iridescent colonies growing from aspirate of ecthyma gangrenosum lesion, plated directly to 5% sheep agar by expelling contents of syringe onto agar surface

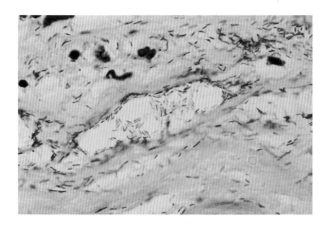

Figure 238 *Pseudomonas aeruginosa* Gram stain of histologic section of ecthyma lesion biopsy, showing numerous bacilli aligned along the inner aspect of intima of blood vessel and within blood vessel. Note also that bacilli have degraded the intima and invaded the surrounding tissue (perivascular cuffing)

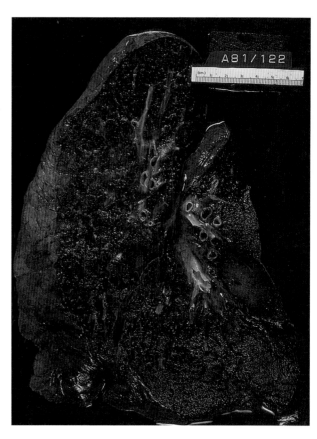

Figure 239 *Pseudomonas aeruginosa* Hemorrhagic pneumonia of lung with targetoid areas of inflamed tissue secondary to vasculitis, caused by invasion of blood vessels by microorganism

ACINETOBACTER SPECIES

Acinetobacter are Gram-negative coccobacilli present in soil and water and on the skin of normal individuals, particularly in moist (axilla, groin) and intertriginous areas. Oral carriage also occurs. This species is second only to *Pseudomonas aeruginosa* as a nonfermentative, nosocomial pathogen, and, analogously, has developed a broad antibiotic resistance pattern. The genus, contained in the family *Neisseriaceae* because of its morphological resemblance to *Neisseria* species, consists of 15 genospecies (DNA hybridization groups), of which two are the most common human isolates *Acinetobacter baumannii*, which oxides glucose, and *A. lwoffi* which is asaccharolytic. In the hospital environment, acinetobacters have been recovered from numerous inanimate sources, e.g. humidifiers, tap water, wash basins, and

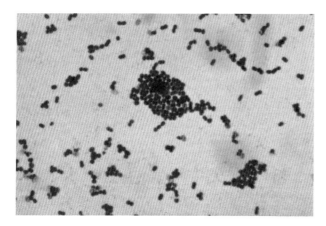

Figure 240 *Acinetobacter baumannii* Diplobacilli and coccal forms in smear of agar culture. Note the tendency of some cells to retain crystal violet

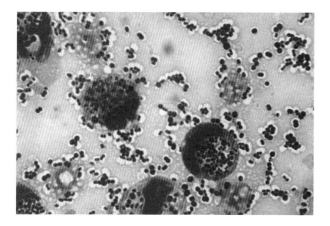

Figure 241 *Acinetobacter baumannii* Intracellular and extracellular encapsulated diplobacilli in peritoneal aspirate of infected white mouse

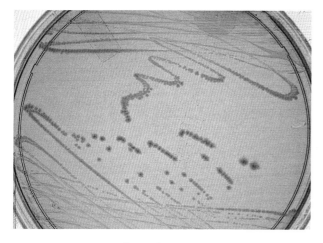

Figure 242 *Acinetobacter baumannii* Smooth, lavender-tinted colonies 24 h after growth on MacConkey agar

ventilatory equipment. *Acinetobacter* species, particularly *A. baumannii*, cause urinary tract and wound infections, bacteremia, and nosocomial pneumonia in intensive care unit patients with severe underlying diseases or having undergone extensive surgical procedures, and who require ventilatory assistance and tracheotomy. Community-acquired pneumonia among elderly patients with underlying disorders also occurs. Post-neurosurgical meningitis is a rare occurrence, as is meningitis in children. Acinetobacter virulence factors include a polysaccharide capsule, adhesiveness to human epithelial cells, and production of exoenzymes, e.g. esterases, mainly by *A. baumannii*.

Morphology

Acinetobacter species are stout, Gram-negative, encapsulated diplobacilli with a tendency to form more elongated forms. In smears of clinical specimens, the diplobacillary form predominates. They are non-motile and oxidase-negative.

Culture characteristics

Acinetobacter grow readily on 5% sheep blood and MacConkey agars, producing smooth, opaque colonies; some isolates are β-hemolytic. Colonies on MacConkey agar are a light lavender color but do not ferment lactose.

BURKHOLDERIA (PSEUDOMONAS) SPECIES

There are three species of note in this genus, the opportunist pathogen *B. cepacia*, and two horse pathogens (glanders) transmissible to humans, *B. mallei* and *B. pseudomallei*, which cause meliodosis, a severe cavitary pulmonary infection with sepsis and metastatic abscesses. These species are of environmental (water, soil, plants) origin; the latter two are found mainly in Asia, Africa, and South America. *B. cepacia* is primarily a nosocomial pathogen surviving in the hospital environment, including in antiseptic solutions, but is also associated with severe necrotizing pulmonary infections in patients with cystic fibrosis and chronic granulomatous disease in childhood. *B. cepacia*, unlike *P. aeruginosa*, can become bacteremic in patients with cystic fibrosis. Endophthalmitis and keratitis have also been described.

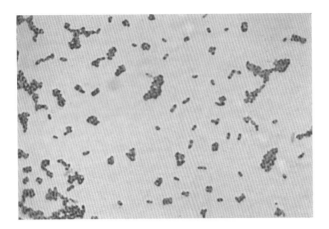

Figure 243 *Burkholderia pseudomallei* Small, slender bacilli with rounded ends and bipolar staining in smear from rough (rugose) colony morphotype. Smooth colony morphotype shows longer, irregularly stained bacilli, often in palisade formation

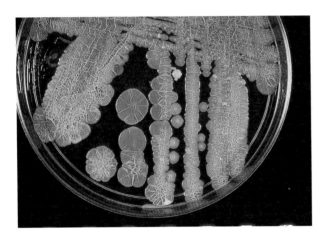

Figure 244 *Burkholderia pseudomallei* Rough (rugose) dry, wrinkled, grayish-yellow striated colonies and smaller, smooth, glistening, yellow-pigmented colonies on nutrient agar

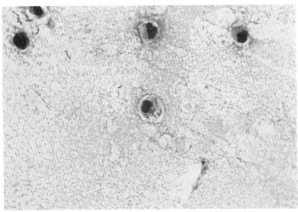

Figure 245 *Stenotrophomonas maltophilia* Slender bacilli with rounded ends singly and within polymorphonuclear leukocyte in smear of aspirate of ecthyma skin lesion in bacteremic patient with acute myelogenous leukemia

Morphology

Burkholderia species are slender, Gram-negative rods with rounded ends. *B. pseudomallei* may also show bipolar staining in smears from infected tissue. *B. cepacia* and *B. pseudomallei* are motile species. In smears of colonies after 48-h growth, *B. pseudomallei* cells may take on an oval morphology with only the periphery of the bacillus staining.

Culture characteristics

Burkholderia species grow readily on bacteriologic media. *B. cepacia* colonies are smooth with a slightly yellow pigment on 5% sheep blood agar. Pigmentation is intensified after growth on Kligler's Iron Agar Slants. *B. pseudomallei* colonies are smooth initially but become dry and wrinkled, with a slight yellow pigment. *B. mallei* colonies are round and smooth with an entire edge. Over time, they may develop a finely granular surface and a grayish-yellow pigment. *B. pseudomallei* is oxidase-positive, whereas the other two species are variable for this trait.

STENOTROPHOMONAS MALTOPHILIA

Formerly classified as *Pseudomonas maltophilia* and *Xanthomonas maltophilia*, this species is a common habitant of soil, plants, water and the hospital environment, having been isolated from thermal humidifiers, antiseptic solutions, and inanimate fomites. In common with *P. aeruginosa* and *Acinetobacter* species, this bacterium can be isolated from respiratory tract infections, bacteremia, urinary tract, and wound infections. Occasionally, bacteremia is accompanied by cutaneous manifestations mimicking pseudomonal ecthyma gangrenosum lesions. Virulence factors include extracellular proteolytic enzymes such as, elastase, hyaluronidase, and a hemolysin. Extracellular DNAse is also produced. Infections occur predominantly in intensive care unit patients with cancers, lymphomas, and leukemia. *S. maltophilia* is also emerging as a pulmonary pathogen in cystic fibrosis patients, assuming the marked encapsulated state often associated with *P. aeruginosa* in this patient population.

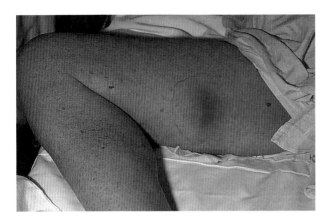

Figure 246 *Stenotrophomonas maltophilia* Raised, indurated ecthyma lesion with duskly violaceous center

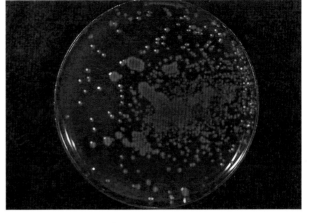

Figure 247 *Stenotrophomonas maltophilia* Well-circumscribed lavender-tinted colonies on 5% sheep blood agar inoculated directly with lesion aspirate. Colonies developed wherever a droplet of aspirate impinged on the agar surface when expelled from a needle and syringe assembly

Morphology

Stenotrophomonas maltophilia is a slender, slightly curved, Gram-negative motile bacillus.

Culture characteristics

Stenotrophomonas grows on most routine bacteriologic media. On 5% sheep blood agar, colonies are slightly raised, smooth, and often tend to become deep green to purple, with a zone of greenish discoloration beneath the colonies. Growth is accompanied by a strong ammonia odor. *S. maltophilia* oxidizes maltose, is oxidase-negative, and hydrolyzes deoxyribonucleic acid.

6

Anaerobes

INTRODUCTION

On the basis of their relationship to oxygen, microorganisms may be classified as *strict aerobes* which require molecular oxygen as a terminal electron acceptor, *microaerophiles* which require reduced oxygen tension and which grow in the presence of 5% CO_2, *facultative anaerobes* which require oxygen under aerobic conditions but can also use organic compounds in a series of oxidation–reduction reactions to grow anaerobically, and *anaerobes* which cannot grow in the presence of molecular oxygen. There is, however, an oxygen gradient within which anaerobic bacteria fall, based upon their capability to neutralize toxic oxygen molecules. Microaerotolerant species produce superoxide dismutase and are able to scavange superoxide anions (O_2^-) which prevent the secondary generation of OH radicals, and singlet oxygen ($^1O_2$), all of which are highly toxic to the bacterial cell. Additionally, many anaerobes involved in human infections, e.g. *Bacteroides* species and *Propionibacterium*, produce catalase- and peroxidase-inactivating hydrogen peroxide (H_2O_2), aborting the generation of O_2^- radicals. Most human anaerobic infections are caused by *microaerotolerant* anaerobes of the genera *Clostridium*, *Peptostreptococcus*, *Fusobacterium*, and *Bacteroides*. Strict anaerobes are either devoid of superoxide dismutase or produce small quantities. Anaerobic bacteria are widely distributed in nature in oxygen-free niches and colonize many sites on the human body including skin (*Proprionibacteriun*). In the colon, the predominant bacterial flora is *anaerobic*, exceeding 10^{11} organisms /gram of colon content and outnumbering facultative species by 1000 : 1. Anaerobes are closely associated with mucosal surfaces in the oral cavity (gingival crevices, tonsilar crypts) and genitourinary tract. In their ecological niche, anaerobic bacteria are generally harmless. When introduced, however, into devitalized tissue with a low redox potential, devastating life-threatening infections may ensue. Usually, infections are polymicrobic and contain aerobic species which diminish oxygen concentrations, thereby facilitating growth of anaerobic species. Anaerobic bacteria have been incriminated in a wide range of human infections, including aspiration pneumonia, pelvic infections, brain abscesses, especially after dental manipulation, intrauterine infection, abdominal infections, subcutaneous infections, bacteremia, and, rarely, endocarditis. Of special note are clostridial gas gangrene, botulism and tetanus. In the oral cavity, two potentially devastating infections can occur, Ludwig's angina and Lemierre's syndrome. Ludwig's angina is a severe, rapidly progressive cellulitis of the floor of the mouth, involving the submandibular and sublingual spaces arising from the second and third submandibular molars. Unimpeded, there is local lymphadenitis, bacteremia, and involvement of the cervical fascia with focal cellulitis. Lemierre's syndrome is an uncommon, but potentially life-threatening entity, that consists of oropharyngeal infection with bacteremia, usually caused by *Fusobacterium mortiferum*. Infection rapidly progresses to jugular vein septic thrombopheblitis, with embolization to the lungs and other tissues. Anaerobic bacteria produce many tissue-destructive exoenzymes, some species-specific.

Acute necrotizing gingivitis, or trench mouth, is an ulcerative process affecting the marginal gingivae, with a propensity to spread to the tonsils or pharynx

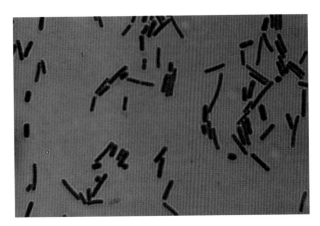

Figure 248 *Clostridium perfringens* Gram stain of culture, showing typical stout bacilli with rounded ends

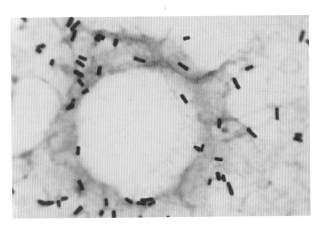

Figure 251 *Clostridium perfringens* Numerous stout bacilli with rounded to square ends in smear of necrotic tissue excised from leg of patient with myonecrosis. Note absence of inflammatory response and presence of clear, circular areas denoting gas bubbles in proteinaceous matrix

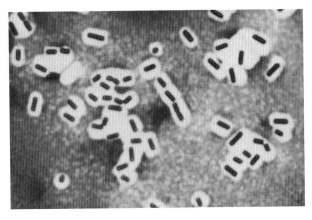

Figure 249 *Clostridium perfringens* India ink preparation of culture smear counterstained with crystal violet, showing distinct capsules. A similar preparation may be used to confirm *C. perfringens* directly in tissue exudates

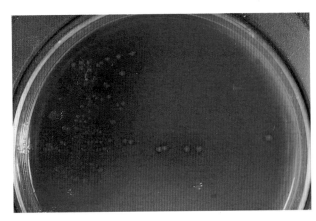

Figure 252 *Clostridium perfringens* Colonies on 5% sheep blood agar after 48-h growth, showing double zone of β-hemolysis

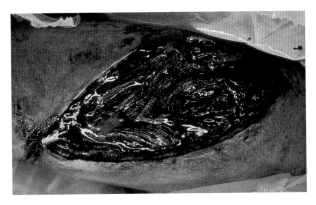

Figure 250 *Clostridium perfringens* Massive myonecrosis of thigh with crepitation extending to the ankle. Tissue destruction is due to local production of toxins, including α-toxin and lecithinase. Gas in tissue occludes blood supply, adding to gangrenous presentation. Patients also have systemic toxemia

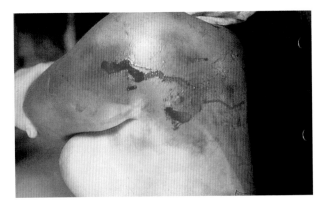

Figure 253 *Clostridium perfringens* Hemorrhagic, painful, oozing bullous lesions on back and shoulder of neutropenic leukemia patient who developed non-traumatic spontaneous cellulitis and myonecrosis. Note the linear creases in skin and rupture of bullous lesions due to pressure of gas formation in tissue. Note also the bluish discoloration of the skin

(Vincent's angina). Spread to involve the facial tissue (cancrum oris, noma) is a major complication. Gram-stained smears of these lesions will show a predominance of Gram-negative cigar-shaped fuso-bacterial organisms and spirochetes (*Borrelia vincenti*).

Necrotizing fasciitis is a rare polymicrobic infection, characterized by rapidly progressive necrosis of deep and superficial fascia and/or muscle, usually originating from, or developing around, the perineal area, with high morbidity and mortality. Radical excisional debridement of infected tissue is germane to treatment.

The clinically significant anaerobes may be divided into the Gram-positive, spore-forming *Clostridium* species and non-spore-forming bacilli *Propionibacterium*, *Bifidobacterium*, *Eubacteriun*, Gram-positive cocci, e.g. *Peptostreptococcus* species, and the Gram-negative rods such as *Bacteroides*, *Fusobacterium*, *Prevotella*, and *Porphyromonas* species.

CLOSTRIDIUM

Clostridia are Gram-positive, spore-forming bacillary species which grow under anaerobic to microaerophilic (*C. perfringens*, *C. septicum*, *C. hemolyticum*, *C. tertium*) conditions. They are widely distributed in soil and comprise part of the normal intestinal flora of humans and animals. Clostridia produce a number of potent exotoxins which characterize their pathogenesis. Among these are the *C. perfringens* α-toxin (hemolysin), *C. tetani* (tetanus) toxin, *C. botulinum* (botulism) neurotoxin, and the *C. difficile* enterotoxin and cytotoxin. *C. botulinum* produces its toxin in food products which are subsequently ingested. In infant botulism, the exotoxin is produced directly in the newborn intestinal tract. Botulinum toxin may also be produced within necrotic tissue (wound botulism) supporting *C. botulinum* growth. Overgrowth of *C. difficile* in the gastrointestinal tract following antibiotic therapy can lead to a toxin-mediated mild diarrhea or a fulminant enterocolitis. Many *Clostridium* species are highly fermentative producing copious amounts of gas (H_2, CO_2) from fermentable substrates. *C. perfringens* is the clostridial species most often recovered from human infection. A large part of its tissue-destructive capacity resides in its rapid generation time of 8 min and production of four potent exotoxins, e.g. α-toxin, proteases, and eight lesser toxins, all of which contribute to

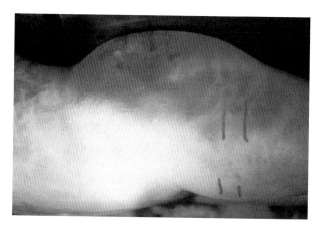

Figure 254 *Clostridium perfringens* Buttocks of patient with spontaneous gangrenous myonecrosis, with bluish discoloration of skin and edema of the surrounding tissue. This entity may also be caused by *Clostridium septicum*

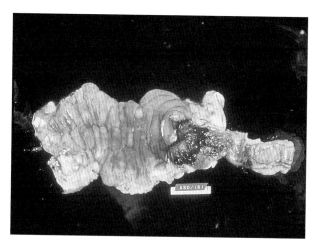

Figure 255 *Clostridium perfringens* Necrotic infarct in bowel of neutropenic leukemia patient who developed spontaneous non-traumatic myonecrosis. Infarcted bowel may serve as the portal of entry of the bacillus into the systemic circulation with focalization in tissue

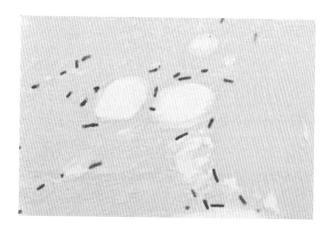

Figure 256 *Clostridium perfringens* Gram stain of histologic section of infected tissue, showing typical non-sporulating bacilli

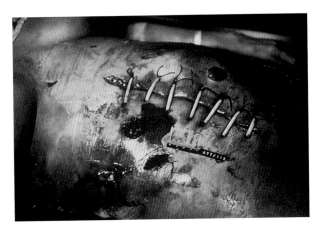

Figure 257 *Clostridium perfringens* Gas gangrene of abdominal wall following intestinal surgery for carcinoma. Note hemorrhagic gas-filled bullous lesions, some which have ruptured, and separation of surgical incision under gas pressure in tissue

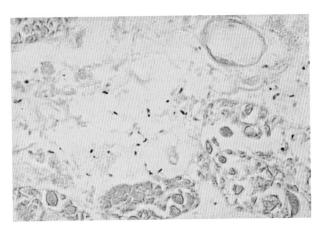

Figure 259 *Clostridium septicum* Gram stain of histologic section of masseter muscle biopsy from neutropenic patient with spontaneous gas gangrene, showing slender bacilli, some ovoid and with spores

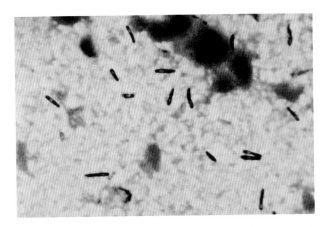

Figure 260 *Clostridium sporogenes* Slightly curved bacilli with parallel sides and rounded ends containing oval spores (clear areas) which slightly bulge into the bacillary body. Smear of aspirate of patient with cellulitis

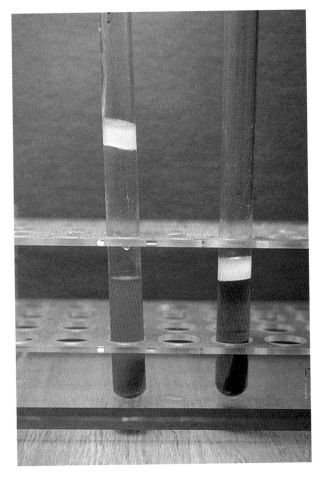

Figure 258 *Clostridium perfringens* Gas production in chopped liver broth culture, as evidenced by elevation of Vaseline plug. Uninoculated control on right

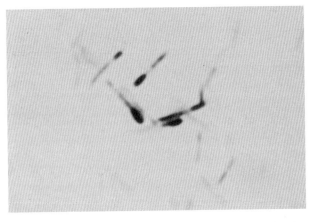

Figure 261 *Clostridium tetani* Characteristic slender bacillus with subterminal spore bulging into bacillary body, imparting a 'drum stick' appearance to the bacillus. The bacillus often decolorizes readily after 48-h growth and stains Gram-negative, as noted in this smear of a thioglycollate broth culture

pathogenesis. On the basis of α-toxin, C. *perfringens* may be divided into five subtypes, A–E, of which type A-producing strains are most common. α-Toxin is a phospholipase which disrupts cell membranes, leading to cell death, and plays a major role in gas gangrene, resulting in muscle necrosis. Other systemic attributes of the toxin include platelet aggregation, increased vascular permeability, and hemolysis. C. *perfringens* type C is found in pigs and is associated with a fulminant necrotizing infection of the small bowel (enteritis necroticans, pig bel) after consumption of undercooked pork containing an enterotoxin (β-toxin)-producing strain. Colonization of the small intestine and overgrowth of C. *perfringens* lead to toxin production and disease. C. *perfringens* type A is associated with a milder form of food poisoning induced by a fluid-inducing heat-labile protein enterotoxin which is part of the structural component of the spore coat and is formed during sporulation.

Morphology

Clostridia present with a range of pleomorphic forms, from stout bacilli with rounded ends to slender bacilli with subterminal spores. While sporulation characterizes all clostridia, C. *perfringens*, one of the common agents of gas gangrene, rarely sporulates in the human body or after growth on bacteriologic media. In soil, however, sporulation ensues readily, enhancing C. *perfringens* contamination and infection of traumatic wounds, e.g. battlefield wounds. Although Gram-positive in young cultures, this staining attribute is readily lost and some species stain Gram-negative after 24–48 h of growth.

Culture characteristics

On agar media, colonies may be discrete or take the form of a fine, spreading film over the agar surface. On 5% sheep blood agar, some species, e.g. C. *perfringens*, produce β-hemolytic colonies.

Clostridium difficile

This species is a slender, Gram-positive bacillus with a large, oval, subterminal spore, and colonizes the human gastrointestinal tract of about 3% of healthy individuals. While it produces a number of extracellular proteases, its overgrowth in the intestinal tract under antibiotic selection leads to the production of two enterotoxins: toxin A, which causes fluid secretion, and toxin B, a potent cytotoxin. Together, these toxins mediate mucosal damage, intestinal fluid secretion, and inflammation. Ultimately, pseudomembranes form, appearing as discrete, raised, white to yellow plaques, which may coalesce as the infectious process continues and stud the entire colon. Diagnosis may be achieved by clinical history of prior antibiotic administration, colonoscopy to detect pseudomembranes, and assay for C. *difficile* toxin in stools. Culture for C. *difficile* on a selective medium, cycloserine cefoxitin fructose agar, which yields yellow colonies, may be undertaken with the proviso that, if recovered, the isolate should be further tested for toxin production. C. *difficile* is a major nosocomial pathogen. Patients with C. *difficile* enterocolitis shed large numbers of spore-bearing bacilli into the hospital environment which can persist for long periods of time and be transferred, mainly on the hands of health-care workers, to other hospitalized patients, leading to colonization or enterocolitis.

BACTEROIDES FRAGILIS GROUP

This group comprises the most frequently encountered species in human infections. Interestingly, although B. *fragilis* comprises a minor part of the gastrointestinal tract flora, it is a primary offender in intra-abdominal infections and bacteremia, which may be associated with encapsulation and exoenzyme production by this species. B. *fragilis* is also catalase- and β-lactamase-positive.

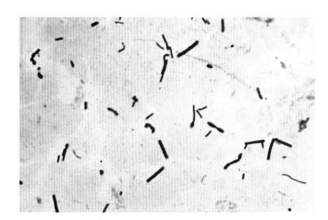

Figure 262 *Clostridium difficile* Gram stain of stool of patient with antibiotic-associated enterocolitis, showing overgrowth of Gram-positive bacilli

Figure 263 *Clostridium difficile* Gray, opaque, non-hemolytic colonies with irregular borders on 5% sheep blood agar. Growth is usually accompanied by a pungent odor

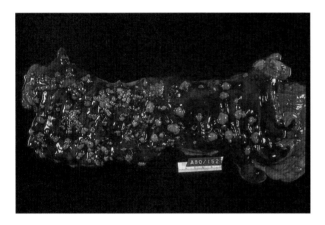

Figure 264 *Clostridium difficile* Section of colon of patient with pseudomembraneous colitis, showing numerous white to yellow raised plaques superimposed on an erythematous, hemorrhagic mucosa

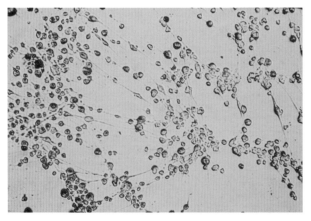

Figure 265 *Clostridium difficile* Positive tissue culture assay for C. *difficile* toxin evidenced by disruption of monolayer and rounding of cells. Depolymerization of microfilaments by toxin results in alteration in cell membrane leading to thread-like projections termed actinomorphic. In presence of toxin-neutralizing antibody (control), the cytopathic effect is eliminated

Morphology

Bacteriodes fragilis is a small, pale-staining, Gram-negative coccobacillus with rounded ends occurring singly or in pairs. In direct smears of clinical material, encapsulation can be observed, imparting to this microorganism a *Haemophilus*-like appearance. *B. ureolyticus* is a thin, delicate, slightly curved rod with rounded ends.

Culture characteristics

Bacteroides fragilis grows well anaerobically on 5% sheep blood agar and on selective laked blood agar containing kanamycin and vancomycin, producing smooth, white to gray, opaque colonies without hemolysis. *B. ureolyticus* colonies on 5% sheep blood agar are rough, highly speckled and pit the agar surface. This colony morphology is similar to that produced by *Eikenella corrodens*, a facultative anaerobe, and from which *B. ureolyticus* must be distinguished as the two species have different antimicrobial susceptibility patterns.

FUSOBACTERIUM SPECIES

Fusobacterium mortiferum is a highly pleomorphic, slender, non-spore-forming bacillus with tapered ends, often containing bulbous swellings which

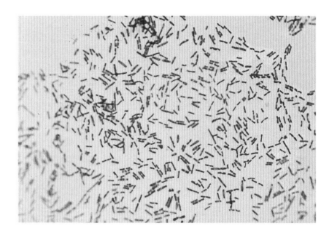

Figure 266 *Bacteroides fragilis* Irregularly staining, straight bacilli with parallel sides and rounded ends in smear of agar culture. Vacuoles and round swelling may also be present. In clinical specimens, small diplobacilli predominate

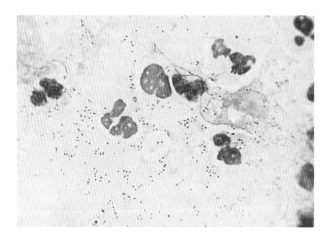

Figure 267 *Bacteroides fragilis* Small, encapsulated cocci singly and in pairs in smear of liver abscess aspirate. Abscess developed secondary to bowel manipulation

Figure 270 *Bacteroides urealyticus* Slender, pale-staining bacilli in smear of exudate from brain abscess in a child

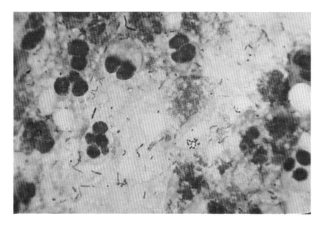

Figure 268 *Bacteroides fragilis* Polymicrobic anaerobic flora including Bacteroides, fusobacteria, and peptostreptococci in smear of brain abscess aspirate

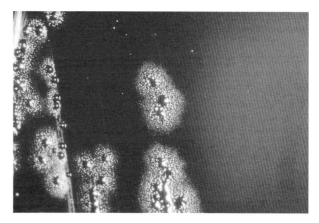

Figure 271 *Bacteroides urealyticus* Highly speckled, slightly spreading colonies with clear central core, as viewed by oblique lighting. Colonies resemble those of *Eikenella corrodens* which is a facultative anaerobe and must be distinguished from *Bacteroides urealyticus*

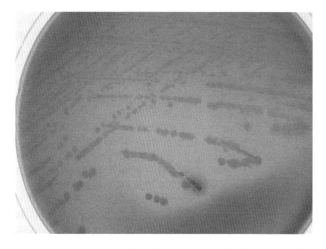

Figure 269 *Bacteroides fragilis* Smooth, shiny, translucent colonies with entire borders on 5% sheep blood agar. Greenish discoloration of medium due to accumulation of metabolites in medium

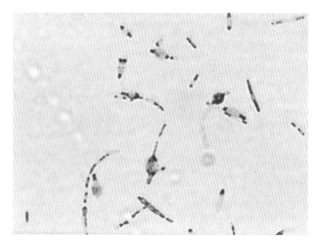

Figure 272 *Fusobacterium mortiferum* Smear of broth culture, showing highly pleomorphic, irregularly staining bacilli in filaments and with large swollen bodies. Unstained clear areas resemble spores

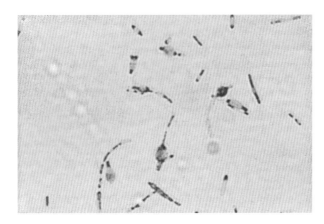

Figure 273 *Fusobacterium necrophorum* Pleomorphic bacilli with pointed ends and bulbous swellings in smear from broth culture

distort the bacillary morphology. Irregular staining gives the impression of spores along the thread forms. *F. necrophorum* may also show bulbous swellings but to a lesser extent. *F. nucleatum* is a long, spindle-shaped bacillus with pointed ends.

Culture characteristics

Depending on the species, colonies on 5% sheep blood agar can be smooth with an entire edge (*F. mortiferum*), circular, raised with green discoloration of the medium (*F. necrophorum*), or rough 'molar tooth'-like with greening of the agar (*F. nucleatum*).

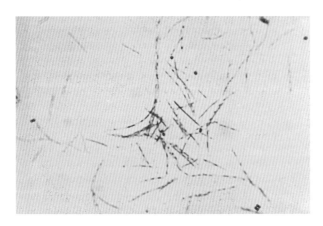

Figure 274 *Fusobacterium nucleatum* Slender, irregularly stained, beaded, spindle-like bacilli with sharply tapered ends in 'stack of hay' alignment

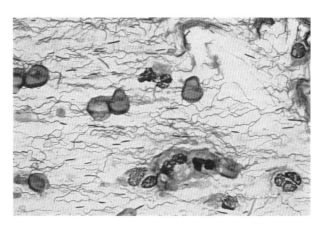

Figure 276 *Fusospirochetosis* Numerous slender, cigar-shaped bacilli with tapered ends and *Borrelia*-type spirochetes in smear of necrotic cheek lesion. Fusospirochetosis represents a synergistic overgrowth of these oral bacteria in the setting of poor oral hygiene ('trench mouth') or pre-existing lesions

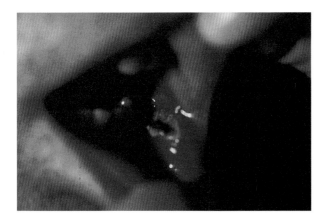

Figure 275 *Fusospirochetosis* Painful ulcer with necrotic membranous base and erythema on inner aspect of cheek in patient with AIDS

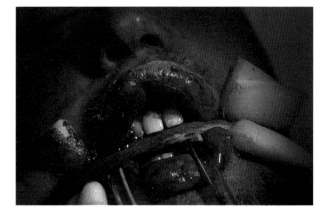

Figure 277 *Fusospirochetosis* Greenish, necrotic, friable ulcerative lesions of the mucosa and gums in an intubated leukemic patient. Scrapings of the mucosa dislodged small granules replete with an overwhelming anaerobic flora, which included fusobacteria and spirochetes

OTHER ANAEROBES

Porphyromonas species and *Prevotella melaninogenica* are Gram-negative coccobacilli characterized by production of colonies with a dark brown to black pigment.

Propionibacterium species are Gram-positive, pleomorphic 'diphtheroid'-like rods with rudimentary branching, often occurring in clusters, especially after growth in liquid media (thioglycollate broth) in which growth takes the form of discrete granules. Colonies on 5% sheep blood agar are smooth and white to tan with age. *Propionibacterium acnes* is catalase-positive.

Bifidobacterium species are Gram-positive, stout bacilli whose ends are clubbed and show short 'forked' bifurcations. Colonies are smooth with irregular contours.

Peptostreptococcus are Gram-positive cocci, with a tendency toward elongation, occurring singly and in chains. Colonies are usually small, opaque or translucent, gray to white.

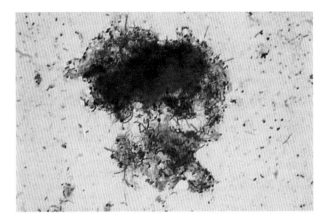

Figure 278 *Fusospirochetosis* Smear of crushed granule revealing dense accretion of polymicrobic anaerobic flora inclusive of long, slender fusobacteria radiating from granule periphery

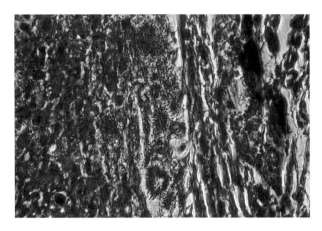

Figure 280 *Fusospirochetosis* Phase-contrast-enhanced microscopy of mucosal biopsy showing packets (colonies) of granules with starburst of fusobacteria from granule periphery. This morphologic presentation of granules can be mistaken for sulfur granules of *Actinomyces*

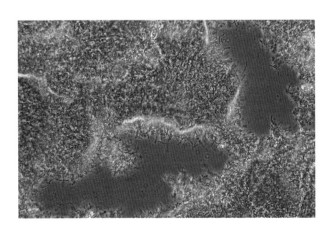

Figure 279 *Fusospirochetosis* Phase-contrast-enhanced microscopy of wet preparation of crushed granule showing enormous microbial flora with peripherally extending slender fusobacteria

Figure 281 *Prevotella species* Black-pigmented colonies on laked blood agar after 72-h incubation under anaerobic conditions

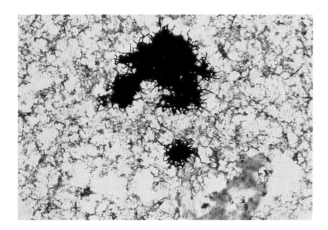

Figure 282 *Propionibacterium acnes* Dense cluster (colony) of pleomorphic bacilli in smear of blood culture broth. Note the striking resemblance to *Actinomyces* species. Granule formation accounts for the 'bread crumb'-like growth in thioglycollate broth

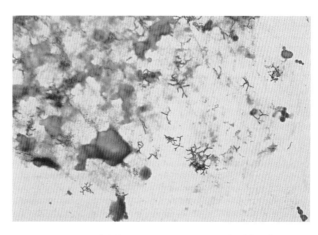

Figure 285 *Bifidobacterium* Species is highly pleomorphic, presenting with irregular length bacilli and filaments, many with terminal swellings and bifurcations, as seen in smear of blood culture

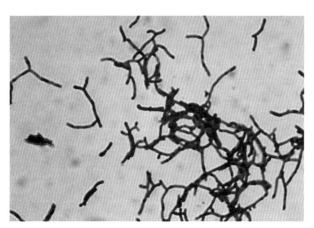

Figure 283 *Propionibacterium acnes* Pleomorphic bacilli with rudimentary branching and swollen ends in smear of agar culture

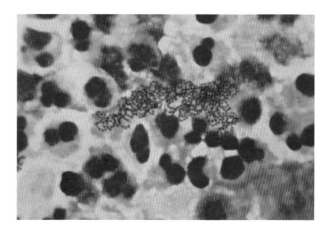

Figure 286 *Peptostreptococcus species* Cocci in interwoven chains in direct smear of brain abscess exudate in patient who had chronic sinusitis which extended to involve the brain

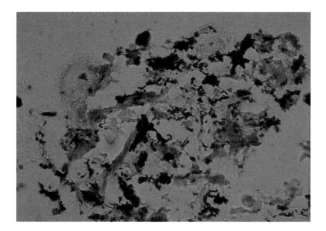

Figure 284 *Propionibacterium acnes* Gram stain of histologic section of brain biopsy of patient with shunt infection showing numerous bacilli singly and in large clusters

7

Fastidious, Gram-negative bacilli

HACEK GROUP

HACEK is an acronym formed by the first letters of a group of bacterial species enjoined not only by their propensity to cause subacute endocarditis, but also for their requirement for increased (5–10%) CO_2 (capnophilic) for growth. Species include *Haemophilus aphrophilus*, *Actinobacillus actinomycetemcomitans*, *Cardiobacterium hominis*, *Eikenella corrodens* and *Kingella* species. *E. corrodens* and *Kingella* species cause a number of other serious infections, including abscesses, osteomeyelitis, and bacteremia. All species are part of the normal oral flora from where they may gain access to tissues, especially after dental manipulation. Definitive virulence factors have not been identified.

Actinobacillus actinomycetemcomitans and *Haemophilus aphrophilus*

These species are considered together because their overall characteristics are similar. *A. actinomycetemcomitans*, as the name implies, is frequently associated with *Actinomyces* species as a feltwork of densely packed, Gram-negative coccobacilli, admixed with the actinomycotic filaments.

Morphology

Both species are small, non-motile, Gram-negative coccobacilli bearing a striking resemblance to *Brucella* species. Longer filament forms may occur.

Culture characteristics

Small, adherent, non-hemolytic colonies develop on 5% sheep blood agar and on chocolate agar.

Continued incubation at 37°C results in a greenish discoloration of the medium surrounding colonies. When viewed by direct microscopy, colonies may contain characteristic, centrally situated 'crossed cigar' or star-shaped configurations. In liquid media, discrete granules (colonies) form along the inner surface of the tube while the broth remains clear.

Cardiobacterium hominis

The species is almost exclusively associated with indolent endocarditis with large vegetations.

Morphology

C. hominis is a pleomorphic rod, originally described as showing swollen ends resembling 'tear drops' especially in smears from liquid media. Smears from

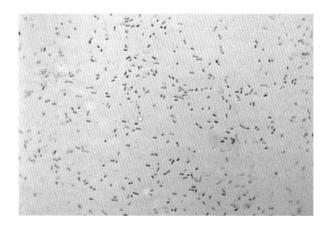

Figure 287   *Haemophilus aphrophilus* Small, pale-staining coccobacilli and bacilli in smear from 5% sheep blood agar culture after 48-h incubation at 37°C

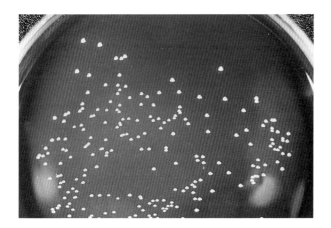

Figure 288 *Haemophilus aphrophilus* White, opaque, sticky colonies with greenish discoloration after 48-h growth on 5% sheep blood agar

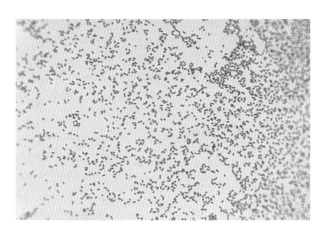

Figure 289 *Actinobacillus actinomycetemcomitans* Small *Brucella*-like coccobacilli seen in smear of 48-h-old 5% sheep blood agar culture. Similar microscopic morphology is noted with *Haemophilus aphrophilus*

Figure 290 *Actinobacillus actinomycetemcomitans* Subculture of positive blood culture showing small sticky colonies with greenish discoloration after 48-h incubation on 5% sheep blood agar in 5% CO_2 at 37°C

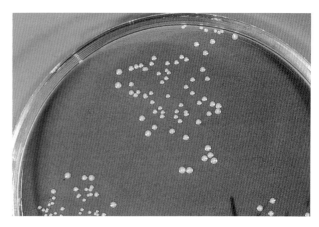

Figure 291 *Actinobacillus actinomycetemcomitans* Small, glistening, sticky colonies after 48-h incubation on chocolate agar in 5% CO_2 at 37°C

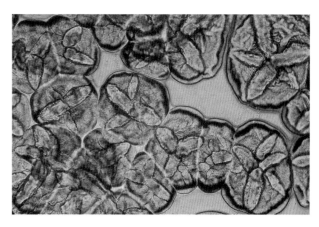

Figure 292 *Actinobacillus actinomycetemcomitans* Presence of characteristic five to six pointed 'crossed cigar' or star-shaped configuration in colonies growing on a clear medium such as Mueller–Hinton agar. Colonies viewed under 100x magnification. This characteristic is shared with *Haemophilus aphrophilus*

Figure 293 *Actinobacillus actinomycetemcomitans* Granular growth adherent to inner surface of glass tube with broth remaining clear. Each granule represents a single colony

agar cultures will show similar morphology but cells arranged in small bundles or 'rosettes' are also observed. There is a tendency for retention of crystal violet, imparting a Gram-variable aspect to the cells.

Culture characteristics

C. hominis grows best under increased CO_2 tension, producing small, round colonies on blood and chocolate agar. A greenish discoloration of the agar medium develops with continued incubation. When viewed microscopically, colonies show a roughened surface, comprised of an intertwinning network of bacilli extending from the colony core to its periphery. C. hominis is indole-positive and fermentative.

Eikenella corrodens

E. corrodens is part of the resident flora of the upper respiratory tract but may also colonize the gastrointestinal and urogenital tracts. No animal or environmental reservoirs exist. Although E. corrodens is regarded as an organism of low virulence, it causes a wide variety of human infections, either alone or in conjunction with various streptococcal species. Trauma associated with human oral flora, e.g. 'clenched fist' injuries or bites, are usually predisposing. Abscesses, endocarditis, meningitis, osteomyelitis, pneumonia and post-surgical infections are among the many *Eikenella* infectious complications.

Morphology

In direct smears of clinical material, E. corrodens appears as coccobacilli resembling *Haemophilus influenzae*, and may even be encapsulated. After growth on agar media, straight bacillary forms with rounded ends ('match sticks') predominate.

Culture characteristics

E. corrodens is facultatively anaerobic (micro-aerophilic) and grows well, albeit slowly, on 5% sheep blood and chocolate agars under increased CO_2 tension, and 35–37°C incubation. Colonies are pinpoint after 24-h incubation but increase in size after 48–72 h, assuming either a characteristic dry, flat, radially spreading, pale yellow-pigmented colony, or a smooth colony morphotype. Colonies may show depressions (pitting) in the agar surface which may be better visualized after the colony

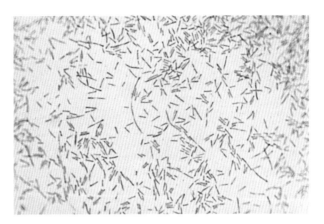

Figure 294 *Cardiobacterium hominis* Slender, pleomorphic bacilli with tapered, slightly, rounded ends in smear of growth on 5% sheep blood agar

Figure 295 *Cardiobacterium hominis* Pleomorphic bacilli, some with swollen ends ('tear drops') and in characteristic clusters of bacilli (rosettes)

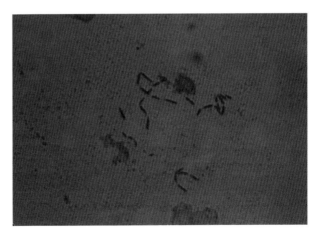

Figure 296 *Cardiobacterium hominis* Smear of positive blood culture of patient with endocarditis on prosthetic heart valve, showing pleomorphic, elongated bacilli with slightly swollen ends

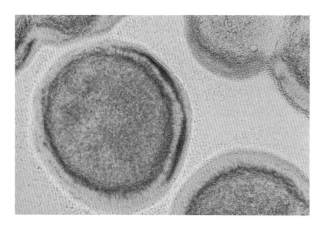

Figure 297 *Cardiobacterium hominis* High-power microscopy of colonies showing rough topography and streaming of bacilli from the colony periphery along the agar surface. This streaming phenomenon is called 'twitching motility'

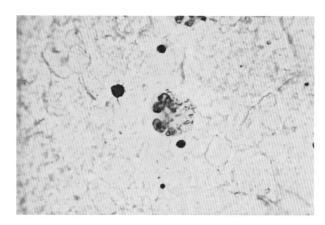

Figure 300 *Eikenella corrodens* Smear of purulent exudate from bite wound to face showing slender rods and diplobacilli within polymorphonuclear leukocyte

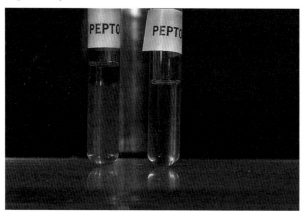

Figure 298 *Cardiobacterium hominis* Positive indole reaction (red color) upon addition of Kovac's reagent to trypticase soy broth supplemented with fetal calf serum to enhance growth. Test was performed after 48-h incubation at 37°C under 5% CO_2

Figure 301 *Eikenella corrodens* Distinctive dry, flat, radially spreading colonies with irregular periphery on 5% sheep blood agar, resembling droplets of condensation fluid

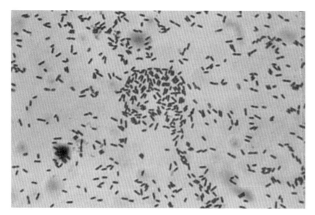

Figure 299 *Eikenella corrodens* Uniformly staining, straight bacilli with parallel sides and rounded ends resembling match sticks. Occasionally, short filaments are also present in smears of agar cultures

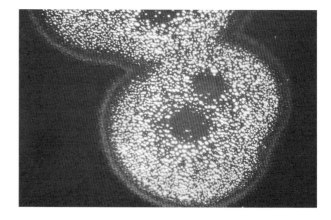

Figure 302 *Eikenella corrodens* Stereomicroscopy of colonies with oblique lighting, showing three distinct zones of colonial growth: a clear, moist, glistening central core, a highly refractile, speckled, pearl-like circle of growth resembling mercury droplets, and an outer perimeter of spreading growth

surface growth has been removed with an inoculating loop. Growth is accompanied by a 'hypochlorite bleach' odor. Growth in liquid media may develop as a uniform turbidity or as discrete granules adhering to the sides of the tube, similar to *A. actinomycetemcomitans*, a feature associated with the hydrophobic fibrillar surface of *E. corrodens* which binds cells together. *E. corrodens* is biochemically inactive and oxidase-positive.

Kingella

The genus, within the family *Neisseriaceae*, is comprised of three species (*K. kingae*, *K. denitrificans*, and *K. indologenes*) which are part of the resident

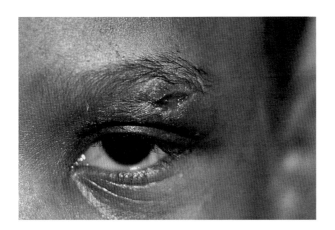

Figure 303 *Eikenella corrodens* Abscess of eye lid in child who was bitten by his brother and developed an abscess and orbital cellulitis requiring incision and drainage

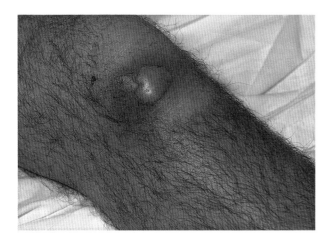

Figure 304 *Eikenella corrodens* Septic arthritis and cellulitis of knee following dental manipulation in a patient with a history of a bleeding tendency and previous injury to his left medial meniscus. An α-hemolytic streptococcus was also isolated from the knee aspirate

flora of the upper respiratory and genitourinary tract. *Kingella* species cause endocarditis and other infectious complications, especially bacteremia, osteomyelitis, and joint infections in children less than 5 years old. Rarely, meningitis has been reported.

Morphology

Kingella species are short, stout, Gram-negative bacilli with rounded or square ends. This morphologic characteristic is shared with *Moraxella* species, as is the tendency to resist decolorization and appear partially Gram-positive. Cells may occur singly or in short chains; often pale-staining 'ghost' forms and large spherical forms are present.

Culture characteristics

Kingella species grow on enriched media such as 5% sheep blood and chocolate agars. On blood agar, *K. kingae* produces two colonial morphotypes, one a spreading, corroding colony associated with 'twitching motility' and fimbriation, the second, a smooth convex colony surrounded by a zone of β-hemolysis. Occasionally *K. denitrificans* has been recovered from Thayer–Martin media used for the isolation of *Neisseria gonorrhoeae* (gonococcus).

BARTONELLA SPECIES

For decades, the genus *Bartonella* contained a single species, *B. bacilliformis*, the causative agent of Carrions' disease, also known as Oroya fever and verruga peruana. Presently, the genus has been expanded to 14 species to include species previously classified as *Rochalimaea quintana* and *Grahamella*. Of these 11 species, four cause major human infections: *B. bacilliformis*, Oroya fever; *B. quintana*, trench fever, bacillary angiomatosis; *B. henselae*, cat scratch disease, bacillary angiomatosis; *B. elizabethae*, endocarditis. *Bartonella* species are thin, slightly curved, Gram-negative rods that stain poorly by the Gram method but can be better visualized by Giemsa and Warthin–Starry silver stains. Epidemiologically, bartonellae have a wide animal distribution and require various vectors for transmission, e.g. *B. bacilliformis* (sandfly), *B. quintana* (scabies mite, body lice), *B. henselae* (cats, fleas, ticks?). Cats are healthy carriers of *B. henselae* and can remain bacteremic for many months or years. Cat-to-cat transmission of the organism is

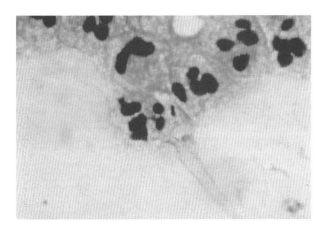

Figure 305 *Kingella species* Plump diplobacilli with square to rounded ends in smear of bone aspirate from a child with osteomyelitis. The morphologic presentation closely resembles that of *Moraxella* species

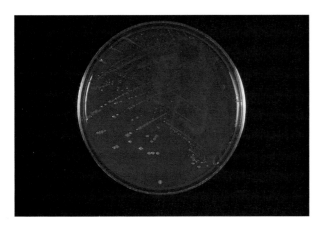

Figure 306 *Kingella species* Small, opaque colonies with even borders after 48-h growth on 5% sheep blood agar. Colonies of *Kingella kingae* would be surrounded by a small zone of β-hemolysis

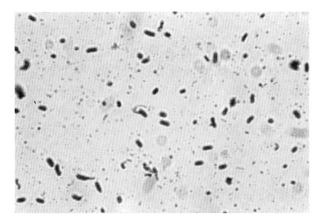

Figure 307 *Bartonella henselae* Pleomorphic bacilli in Warthin–Starry silver stain of agar culture. Tail-like appendages are coalesced pilli (courtesy of Philip M. Tierno, PhD)

mediated by the cat flea. *Bartonella* bacteremia in animals peaks during late summer, coincident with heavy flea infestations and the incidence of cat scratch disease. During infection of their animal reservoir or human host, these arthropod-borne pathogens typically invade and persistently colonize mature erythrocytes. In both hosts, however, endothelial cells are the preferred target cells for bartonellae.

B. henselae and *B. quintana* are found world-wide; *B. bacilliformis* occurs along the western slopes of the Andes Mountains in Peru, Columbia, Ecuador, Chile, Bolivia, and Guatamala. *B. bacilliformis* produces a biphasic disease consisting of a life-threatening febrile anemia phase and, in survivors, a secondary phase characterized by vasoproliferative (angiomatosis) eruptions caused by growth of the microorganism in the vasvular (verruga) epithelium of the skin. *B. quintana*, the causative agent of trench fever, is characterized by recurrent cycling fever and is transmitted amongst humans by the human body louse, *Pediculus humanus corporis*. Outbreaks referred to as urban trench fever have occurred in alcoholic homeless individuals who had lice or scabies. *B. henselae* is most closely linked to cat scratch disease, which presents in normal individuals as prolonged self-limiting regional lymphadenopathy. The disease usually begins with a primary cutaneous inoculation lesion at the site of a cat scratch or bite. *B. henselae* has also been associated with bacteremia in immune-impaired persons, giving rise to cutaneous bacillary hemangiomatosis and bacillary peliosis. Bacillary hemangiomatosis commonly presents with multiple subcutaneous blood-filled nodules, consisting of proliferating endothelial cells; lesions may also be present in bone, lung, central nervous system, liver, spleen, lymph nodes, conjunctiva, and mucosal surfaces of the gastrointestinal and respiratory tracts. Lesions in liver or spleen are referred to as bacillary peliosis hepatis or bacillary peliosis splenitis. Angiogenic cutaneous lesions are absent in normal individuals with cat scratch disease or trench fever.

AFIPIA FELIS

In 1988, the Armed Forces Institute of Pathology, using the Warthin–Starry silver stain, identified a bacterium in the lymph nodes of patients diagnosed with cat scratch disease which was subsequently isolated and given the name *Afipia* after its site of recognition and later the species designation *felis* because of the presumed association with cats. Although thought to be the agent of cat scratch disease, no convincing serologic or cellular immune response to *A. felis*-derived antigens has been forthcoming among patients with cat scratch disease, and, unlike patients with *B. henselae* infection, no clear linkage to cats has been demonstrated.

Morphology

Both *B. henselae* and *A. felis* are slender, poorly staining, small, Gram-negative bacilli. *B. henselae* is slightly curved, bearing a morphologic resemblance to *Campylobacter* species. *A. felis* is motile through a single polar or lateral flagellum, whereas *B. henselae* is non-motile.

Culture characteristics

B. henselae grows best on fresh chocolate agar or heart infusion agar containing horse or rabbit blood, with incubation at 35–37°C, under increased CO_2, and high (80%) humidity. Colonies, which may require more than a month to develop, are white to buff, dry, and adherent (embedded) to the agar surface. *A. felis* grows best at 32°C incubation on buffered yeast extract agar (used for *Legionella* isolation) and in nutrient broth. Colonies are gray-white, glistening, raised, and opaque.

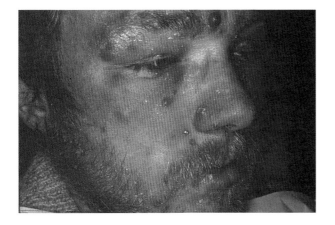

Figure 308 *Bartonella henselae* Bacillary angiomatosis manifested as discrete blood-filled violaceous papules on face of patient with AIDS and cat scratch disease. Note also the butterfly rash of seborrheic dermatitis, caused by an overgrowth of the lipophilic skin fungus *Malassezia furfur* (*Pityrosporum orbiculare*) (courtesy of Donald Rudikoff, MD)

BORDETELLA SPECIES

The genus *Bordetella* is comprised of tiny coccobacilli, of which two species, *B. pertussis* and *B. parapertussis*, are colonizers of the human respiratory tract and cause pertussis or 'whooping cough'. A third species, *B. bronchiseptica*, is a respiratory pathogen of animals, especially rodents, and rarely causes human infections. Whooping cough is a toxigenic, non-invasive infection which occurs mainly in non-immunized individuals and which is characterized by severe coughing (paroxysmal phase), lasting for many weeks if left untreated. During paroxysmal debilitating coughing, there is no pause for air intake until a final cough that clears tenacious airway secretions and is followed by an inspiratory burst through a narrowed glottic opening, producing a characteristic high-pitched whoop. Between paroxysms, individuals appear well, without coughing or fever. Infection begins by colonization of ciliated epithelium of the trachea and bronchi, prolific multiplication of the micro-organism, and production of a cilia-paralyzing tracheal cytotoxin (among other toxins), which disrupts normal clearing mechanisms. Local tissue damage and sloughing of respiratory mucosal epithelium occur. Pertussis is a multifactorial disease and not solely related to toxin production, as exemplified by the fact that *B. parapertussis*, which is non-toxigenic, produces a similar, although milder disease. Immunization inhibits attachment of *B. pertussis* to respiratory tract epithelia.

Morphology

Bordetella species are tiny, Gram-negative coccobacilli resembling *Brucella* and *Haemophilus* species. *B. bronchiseptica* is highly motile by peritrichous flagella.

Culture characteristics

B. pertussis and *B. parapertussis* are not dependent on X (hemin) or V (nicotinamide adenine dinucleotide) factors for growth, but do require a complex medium containing sheep blood, potato extract, and glycerol for growth (Bordet–Gengou medium). Shiny, smooth, pearl-like colonies with an entire edge develop after 48-h incubation under increased moisture. A charcol agar supplemented with 10% horse blood and 40 mg/l of the antibiotic cephalexin has also been formulated. The most commonly used

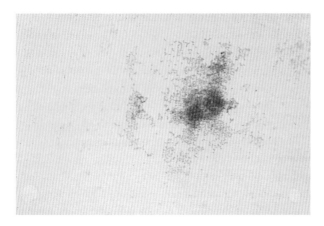

Figure 309 *Bordetella pertussis* Minute, pale-staining coccobacilli in smear from agar culture. Some longer bacillary forms are also present

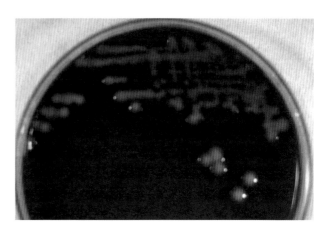

Figure 312 *Bordetella bronchiseptica* Smooth, mucoid, opaque, coalescing colonies on 5% sheep blood agar after 72-h incubation. Colonies, which were initially pinpoint after 24-h incubation, were derived from subculture of a positive blood culture from a patient with AIDS and indwelling Broviac catheter. Colonies also developed on MacConkey agar. The patient had recurrent *B. bronchiseptica* bacteremia which did not clear until catheter removal

Figure 310 *Bordetella pertussis* Small, silvery ('pearl-like') colonies on Bordet–Gengou agar after 72-h incubation in a moist atmosphere

version of this medium is Regan–Lowe. *B. parapertussis* produces a brown, diffusible pigment on nutrient agar. *B. brochiseptica* grows readily on most media including MacConkey agar within 24 h.

BRUCELLA

Brucella species are primarily pathogens of animals, affecting their reproductive organs and causing sterility or abortions. Humans are infected directly by handling infected animals or indirectly through consuming contaminated food products, e.g. unpasteurized milk, cheese, etc. Infection rates are particularly high in shepherds, abattoir and dairy workers, butchers, veterinarians and in microbiologists who may acquire the microorganism from cultures, either by splattering, or through inhalation of aerosols while 'sniffing' unidentified cultures. The genus contains six species of which *B. abortus, B. canis, B. melitensis* and *B. suis* infect humans. Three of these species have primary animal hosts: *B. abortus* (cattle), *B. canis* (dogs), *B. suis* (pigs). *B. melitensis* is the most virulent species and its zoonotic hosts are goats, sheep and cattle. It is the most common species recovered from human infections. Brucellae are facultative intracellular pathogens. They can survive and multiply in cells of the reticuloendothelial tissue by inhibiting the bactericidal

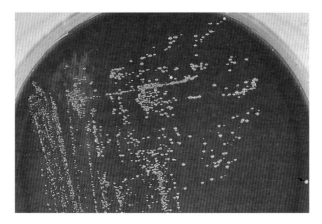

Figure 311 *Bordetella pertussis* Smooth, opaque colonies developing on 5% sheep blood 3–5 days post-subculture from Bordet–Gengou agar

activity of the myeloperoxidase–peroxide halide system of mononuclear phagocytes, and inactivating superoxide anions through superoxide dismutase production. Brucellosis is a bacteremic disease which involves many organs and tissue of the body, accompanied by a chronic, relapsing 'fever of unknown origin' (undulant fever). Complications include endocarditis and osteomyelitis.

Morphology

Brucellae are small, Gram-negative, coccobacilli that are non-motile and non-encapsulated. In smears from agar media, the microorganism is so small and coccal as to make bacillary morphology difficult to discern. In liquid media, short chains form.

Culture characteristics

Brucella species are strict aerobes but grow better under increased (5–10%) CO_2 tension, producing on 5% sheep blood or chocolate agars small, smooth, non-hemolytic colonies after 48-h incubation. As *Brucella* species are highly infectious, unidentified Gram-negative, small coccal isolates can be screened by urease testing on Christensen's medium; if positive within 2 h, *Brucella* should be suspected and cultures handled with great care under biohazard conditions.

CAPNOCYTOPHAGA

Capnocytophaga species may be divided into those which colonize the human oral cavity (C. *ochracea*, C. *gingivalis*, C. *sputigena*), which are slender, Gram-negative, fusiform-like bacilli, and those carried in the oral cavity of dogs (C. *canimorsus*, C. *cynodegmi*), which are highly pleomorphic. The former species play a role in juvenile periodontitis and occasionally cause a bacteremia in neutropenic and immuno-compromised patients. Eye infections, such as chronic conjunctivitis and keratitis, have also been described, as well as rare association with sexually transmitted disease. These oral species produce a number of tissue-destructive exoenzymes, including a leukotoxin. C. *canimorsus*, more so than C. *cynodegmi*, can cause fatal overwhelming bacteremia and septic shock, leading to disseminated intra-vascular coagulation subsequent to a dog bite. This syndrome occurs most frequently, but not exclusively, in patients who may have liver disease,

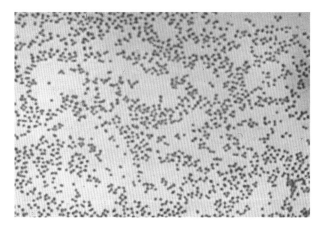

Figure 313 *Brucella melitensis* Small, oval cocci and coccobacilli in smear of growth on 5% sheep blood agar. *Brucella* species frequently occur in coccoid morphology rather than bacillary form. The photograph is slightly enlarged

Figure 314 *Brucella melitensis* Subculture of positive blood culture to chocolate agar, resulting in small, opaque, smooth colonies after 48-h incubation in 5% CO_2

Figure 315 *Brucella melitensis* Author 'sniffing' uniden-tified culture, allegedly leading to inhalation of *Brucella* microorganism and development of brucellosis. Infection required hospitalization for 3 days and antibiotic therapy for 6 weeks to resolve

Figure 316 *Capnoctophaga species* Long, slender, beaded, fusiform-like bacilli with tapered ends after 48-h growth under 5% CO_2

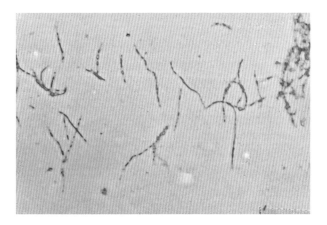

Figure 317 *Capnoctophaga species* Phase-contrast microscopy of Gram stain of culture highlighting slender, beaded filaments with tapered ends

be neutropenic, or have been splenectomized. The lipopolysaccharide (LPS) endotoxin moiety of *C. canimorsus* is highly active.

Morphology

Oral *Capnocytophaga* species are characterized by their long, delicate, filamentous threads with pointed ends. These species are motile with a thin, polar flagellum and exhibit gliding motility on agar media. *C. canimorsus* is a highly pleomorphic, spindle-shaped, filamentous bacterium often displaying round, bulbous swelling. Staining may be irregular.

Culture characteristics

Capnocytophaga require increased CO_2 tension for growth and produce colonies with irregular borders, which have a tendency to spread, often pulling ('piggy back') other colony forms with them. Colonies are non-hemolytic and show a yellow pigmentation, especially on a cotton swab passed thorough the colony. Colonies may pit the agar. In addition to growing on 5% sheep blood agar and chocolate agar, growth may also occur on Thayer–Martin medium used for the recovery of *Neisseria gonorrhoeae*.

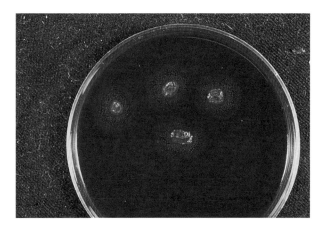

Figure 318 *Capnoctophaga species* Radially spreading, speckled, silver iridescent colonies from point inoculation. Colony transmigration is due to 'twitching motility' of bacilli from the leading edge of the colony

Figure 319 *Capnocytophaga species* Colonies on 5% sheep blood agar showing raised silvery central core with molar tooth texture and spreading borders

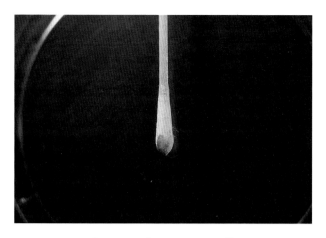

Figure 320 *Capnocytophaga species* Yellow pigment on cotton swab after passage through colonies

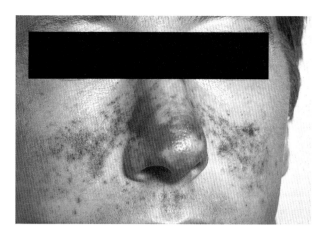

Figure 323 *Capnocytophaga canimorsus* Purpuric, violaceous rash and gangrene of tip of nose in patient who sustained dog bite and developed fatal disseminated intravascular coagulation (courtesy Burt R. Meyers, MD)

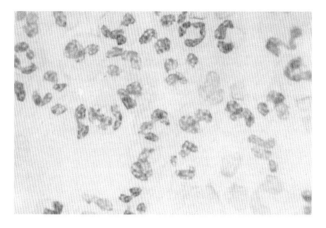

Figure 321 *Capnocytophaga species* Faint staining of slender bacilli within leukocyte in smear of conjunctiva from patient with chronic blepharoconjunctivitis. This species is underappreciated as a cause of conjunctivitis, canaliculitis, endophthalmitis, and keratitis

FRANCISCELLA TULARENSIS

The genus *Franciscella* is composed of minute, Gram-negative coccobacilli with two species, *F. tularensis* with biogroups *tularensis*, *novicida*, and *palaearctica*, and *F. philomiragia* (formerly *Yersinia philomiragia*). *F. tularensis* is the causative agent of tularemia, a human zoonosis, contracted from infected rodents, ground squirrels, hares and jack rabbits by handling them directly, or indirectly through the bites of ticks or deer flies. In the United States, disease occurs in the southern and southwestern part of the country (Missouri, Oklahoma, Texas, Kansas). Well-water contaminated by infected rodents may also serve as a source for the microorganism. *F. tularnensis* biogroup *tularensis* is the most virulent species; *F. tularensis* biovar *novicida* and *F. philomiragia* are minor pathogens. The clinical manifestations of tularemia and their categorization into glandular, uceroglandular, oculoglandular, oropharyngeal, typhoidal and pneumonic depend on route of acquisition or portal of entry of the microorganism. *F. tularensis*, analagous to *Brucella* species, is highly infectious, requiring as few as ten biotype *tularensis* inoculated subcutaneously, or 25 aerosolized cells to establish infection. When ingested in contaminated food, 10^8 organisms are infectious. *F. tularensis* is a facultative intracellular pathogen surviving for prolonged periods in macrophages by inhibiting

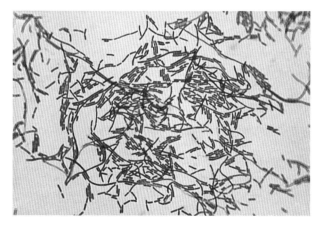

Figure 322 *Capnocytophaga canimorsus* Characteristic irregular-staining filamentous forms with large, spherical swellings in smear from agar culture

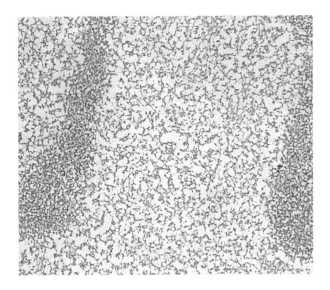

Figure 324 *Franciscella tularensis* Tiny coccobacilli and bacilli in smear of agar culture (courtesy of J. Michael Janda, PhD)

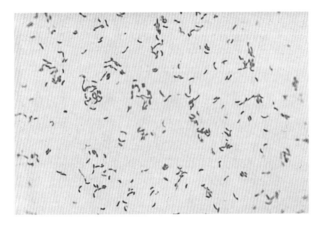

Figure 325 *Gardnerella vaginalis* Slightly curved bacilli, some with terminal clubbing, arranged in small parallel groupings in smear from agar culture

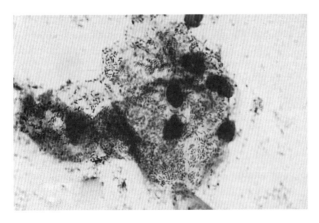

Figure 326 *Gardnerella vaginalis* Vaginal squamous epithelial cells ('clue' cell) containing numerous adherent, Gram-negative bacilli. There are also many bacilli which are unattached

phagosome–lysosome fusion. Additionally, *F. tularensis* possesses a lipid-rich antiphagocytic capsule.

Morphology

F. tularensis is a minute, non-motile, non-spore-forming, pale-staining, Gram-negative coccobacillus. Capsules may be discerned in clinical material. In young cultures, marked pleomorphism, manifested by swollen oval bodies, may be seen. Bipolar staining is best demonstrated by Giemsa stain.

Culture characteristics

It should be noted that, as with *Brucella* species, culture isolation in the microbiology laboratory is particularly hazardous and requires a biosafety cabinet and protective equipment for handling specimens. *F. tularensis* is a fastidious aerobe requiring media supplemented with cystine or cysteine and defibrinated rabbit or human packed red blood cells. Growth may also be obtained on chocolate agar enriched with Isovitalex. Colonies, which develop slowly, are pale, translucent, and slightly mucoid.

GARDNERELLA VAGINALIS

G. vaginalis is a natural colonizer (30–40%) of the human vagina and is frequently associated with bacterial vaginosis, a massive overgrowth of vaginal indigenous flora including *G. vaginalis*, anaerobes, *Mobiluncus* species, and the genital mycoplasmas, *Ureaplasma urealyticum* and *Mycoplasma hominis*. While no single bacterial species is etiologically related to bacterial vaginosis, *G. vaginalis* may be found in 50–60% of patients with the entity. Bacterial overgrowth may result when the normal *Lactobacillus* flora of the vagina is altered. Lactobacilli inhibit growth of other bacteria by production of lactic acid (which maintains vaginal pH < 4.5), hydrogen peroxide, and an inhibitory protein (bacteriocin). Bacterial overgrowth is accompanied by vaginal leukorrhea and an increased production of amines (trimethylamine) by anaerobic species, which renders a 'fishy' odor when vaginal secretions are emulsified in 10% potassium hydroxide. The combined attributes of bacterial overgrowth and abnormal amine production lead to a sloughing of vaginal squamous epithelial cells to which *G. vaginalis* avidly adheres, forming the characteristic 'clue' cell. The absence of lactobacilli in either wet preparation or Gram stain is also of diagnostic importance. In addition to association with bacterial vaginosis, *G. vaginalis* also causes urinary tract infections in female and male patients, amnionitis, and bacteremia in both male and female patients.

Morphology

G. vaginalis is a pleomorphic, small, Gram-negative to Gram-variable rod. Cells may occur in angular and palisade arrangement, resembling *Corynebacterium* species. Dark-staining granules may also be present. By cell wall ultrastructure and chemical composition, *G. vaginalis* is a Gram-positive bacterium, although it stains Gram-negative to Gram-variable.

Culture characteristics

G. vaginalis produces small, opaque, β-hemolytic colonies on human blood but not sheep blood agar. A selective medium designated HBT (human-blood-bilayer-tween) enhances isolation and hemolytic activity.

HAEMOPHILUS

The genus *Haemophilus* is contained in the family *Pasteurellaceae* and contains eight species of small, Gram-negative coccobacilli, of which *H. influenzae*, *H. parainfluenzae*, and, to a lesser extent, *H. aphrophilus* (discussed under HACEK organisms) and *H. ducreyi*, the causative agent of chancroid, are the more common isolates. The genus is defined by the fastidious growth factor requirements of its members, particularly hemin (X factor) and nicotinamide adenine dinucleotide (V factor). Growth is further enhanced by increased CO_2 tension. X and V factor requirements play a major role in identifying *Haemophilus* species. *H. influenzae* is strictly human-bound, colonizing mainly mucous membranes of the throat and nasopharynx; isolates have also been recovered from the genital and gastrointestinal tracts. Carriage rates in the respiratory tract for *H. parainfluenzae* and non-encapsulated *H. influenzae* range from 25 to 75%, whereas, for encapsulated, and hence more invasive *H. influenzae*, e.g. *H. influenzae* type B, carriage rates in normal individuals are from 3 to 5%. There are six serotypes (A–F) of

Figure 327 *Haemophilus influenzae* Pale-staining, minute, almost coccal, coccobacilli and a few longer bacillary forms in smear of chocolate agar culture

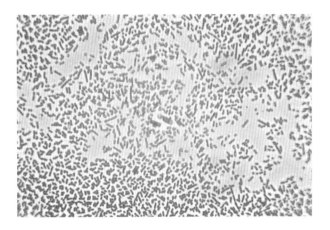

Figure 328 *Haemophilus influenzae* Phase-contrast microscopy of Gram stain, highlighting range of pleomorphic forms, with a predominance of bacilli, some with oval bodies

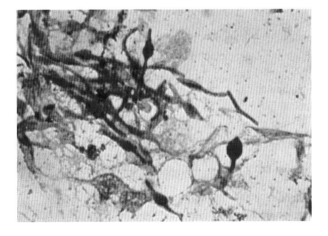

Figure 329 *Haemophilus influenzae* Long, thread-like bacilli with enormous fusiform swelling at end of filament in smear of blood culture of patient being treated with a cell wall-active β-lactam antibiotic. This morphologic presentation may also be encountered in smears in the absence of an inducing agent

encapsulated *H. influenzae*, of which serotype B causes most of the invasive systemic infections, including bacteremia, meningitis, bone and tissue infections, pneumonia, and acute epiglottitis, which can cause airway obstruction and fatality. In the past, *H. influenzae* type B meningitis principally affected children aged 2 months to 2 years, but the incidence of this infectious complication has significantly decreased since the introduction of the *H. influenzae* type B vaccine in industrialized countries. Non-encapsulated *H. influenzae* are involved in otitis media, acute sinusitis and respiratory tract infections in individuals with chronic bronchitis, and as a cause of community-acquired pneumonia. *H. parainfluenzae* has been particularly associated with endocardi-

tis and incriminated in acute epiglottis in adults. *H. influenzae* biotype *aegyptius* caused an outbreak in Brazil of purulent conjunctivitis in children 1–4 years of age, followed by acute onset of fever, bacteremia, petechiae, purpura, vascular collapse, hypotensive shock and death, usually within 48 h of onset. This outbreak, termed Brazilian purpuric fever, was caused by a single clone of *H. influenzae* biogroup *aegyptius*. *H. influenzae* is transmitted primarily by respiratory secretions. Colonization of mucous membranes is enhanced by fimbrial adhesins and IgA protease. Encapsulation retards phagocytosis and complement-mediated serum bactericidal activity.

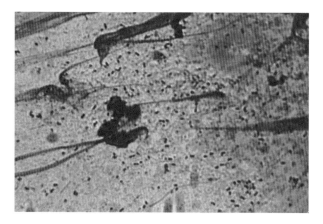

Figure 330 *Haemophilus influenzae* Encapsulated, short coccal forms and coccobacilli in direct smear of purulent sinus drainage of patient with chronic sinusitis

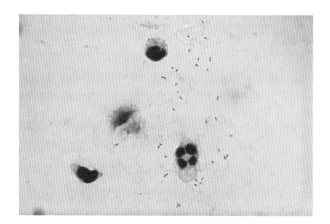

Figure 331 *Haemophilus influenzae* Small coccobacilli and bacilli in direct smear of cerebrospinal fluid of child with *Haemophilus influenzae* type B meningitis. Introduction of type B vaccine has dramatically reduced the incidence of this infectious complication in infants less than 2 years of age

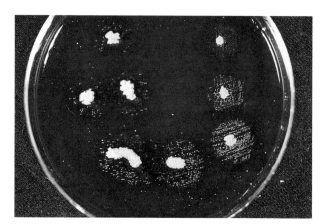

Figure 332 *Haemophilus influenzae* Satelliting growth around hemolytic colonies of *Staphylococcus aureus*. Hemolysis of red blood cells liberates intracellular nicotinamide adenine dinucleotide (V factor) which, in conjunction with hemin (X factor) in blood agar, promotes growth of *H. influenzae*. Satelliting colonies are delineated by the gradient of diffusion of the staphylococcal hemolysin. *Haemophilus parainfluenzae*, which requires only V factor, will render the same satellite phenomenon on blood agar. The test should be performed on a non-blood-containing medium using a staphylococcal isolate or commercially prepared X and V factor-impregnated strips or discs to determine the need for V factor alone

Figure 333 *Haemophilus influenzae* Smooth, translucent, slightly mucoid colonies on chocolate agar. Growth is accompanied by a musty 'mousy odor'

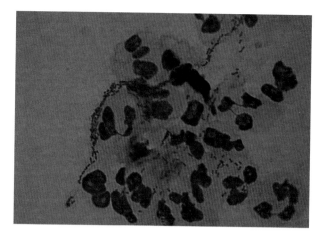

Figure 334 *Haemophilus ducreyi* Short, compact, bipolar-staining bacilli with rounded ends, occurring singly, in pairs, and in long chains of parallel bacilli ('school of fish'). Intercellular adhesion accounts for the parallel rows of bacilli

Morphology

H. influenzae is a small, Gram-negative, encapsulated coccobacillus to short bacillus. In smears of respiratory tract specimens, the microorganism may be obscured in the proteinaceous background because of its small size. Examination of polymorphonuclear leukocytes in the exudate may reveal intracellular coccobacilli, which will aid detection.

Culture characteristics

Haemophilus species are facultative anaerobes which grow readily (some species slowly) on culture media supplemented with growth factors. On 5% sheep blood agar, *H. influenzae* will only grow around colonies of another microorganism (Satellite phenomenon), e.g. *Staphyloccus aureus*, which produces a hemolysin lysing the erythrocytes, thereby liberating intracellular nicotinamide adenine dinucleotide (V factor). Satelliting colonies are often pinpoint and best observed by tilting the Petri dish. On chocolate agar prepared by heat-lysing erythrocytes to liberate V factor, colonies of encapsulated *H. influenzae* are smooth and opaque after 24-h incubation at 37°C under 5–10% CO_2. *H. influenzae* is X and V factor-dependent, whereas *H. parainfluenzae* requires only V factor.

Haemophilus ducreyi

This species is the etiologic agent of chancroid or soft chancre, a sexually transmitted disease characterized by a well-circumscribed, painful penile or labial ulcer with ragged edges and a granular appearance. Infection is often accompanied by painful, tender, unilateral lymphadenopathy with fluctuance termed a chancroidal bubo which may spontaneously rupture, discharging greenish-yellow pus. *Haemophilus ducreyi* is only carried by humans and it gains entry through breaks in the skin. Transmission is person to person; satellite ('kissing') lesions are common.

Morphology

In direct smears of ulcers best prepared by pressing a glass slide directly onto the lesion, *H. ducreyi* will appear as a slender, Gram-negative, irregularly stained bacillus, occurring singly, in masses, intracellular in leukocytes or extracellular. Bacilli may also be seen in long, parallel rows, 'school of fish' arrangement, coursing between inflammatory cells. This phenomenon may be associated with the adhesiveness of the *H. ducreyi* outer membrane.

Culture characteristics

H. ducreyi is a fastidious microorganism growing best on enriched media such as chocolate agar supplemented with isovitalex (yeast extract, amino acids), and in 5–10% CO_2, increased humidity and incubation at 33–35°C. Colonies are round, raised, compact with a granular texture and extremely cohesive, enabling them to be dragged along the agar surface intact. A tan to yellowish pigment may develop after 72-h incubation.

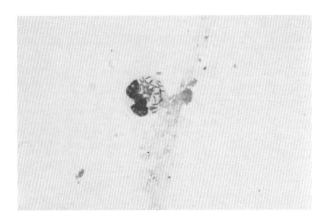

Figure 335 *Haemophilus ducreyi* Short bacilli within polymorphonuclear leukocyte in touch imprint of penile lesion

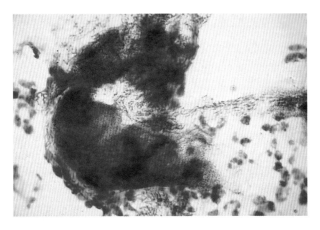

Figure 336 *Haemophilus ducreyi* Dense 'school of fish' arrangement of bacilli entwined with a mucus strand and inflammatory exudate

Figure 337 *Haemophilus ducreyi* Raised, buff-colored, opaque, compact, small and large colonies on Thayer–Martin chocolate agar after 48-h incubation in 5–10% CO_2 and a humid atmosphere. Variation in colony size, shape, and opacity gives the impression of a mixed culture. Because of their compact nature and intercellular binding of cells, colonies may be pushed across the agar surface with an inoculating loop

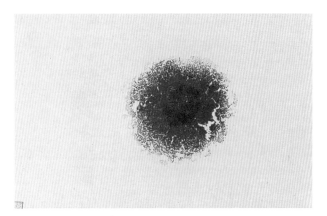

Figure 338 *Haemophilus ducreyi* Low-power microscopy of smear of colony showing dense center packed with bacilli and less dense colony periphery, imparting a 'fried egg' appearance to colony

LEGIONELLA

The single genus in the family *Legionellaceae* contains 39 species and numerous serogroups, of which *L. pneumophila*, serotypes 1, 4, and 6, account for the majority of human infections. Legionellae had been largely overlooked as a cause of epidemic and sporadic cases of pulmonary infections, because of their imperceptible staining in clinical specimens by the Gram method, and their exacting nutritional requirements, e.g. L-cysteine and iron salts for growth on bacteriologic media. Legionellae are slender, Gram-negative, aerobic, motile bacilli with a world-wide distribution in water sources in natural environments (streams, rivers, lakes, etc.) and inanimate surroundings, such as air conditioning, cooling towers, and showerheads, in which they may survive for prolonged periods of time. Legionellae can also survive in the presence of elevated temperatures and chlorination. *L. pneumophilia* mainly cause two pulmonary syndromes: a relatively benign, febrile illness without pneumonia (Pontiac fever), and a more acute, multi-organ disease affecting the lungs (pneumonia), gastrointestinal tract, liver, kidneys and central nervous system (Legionnaires' diseases). Severity of Legionnaires' disease ranges from a mild infection to multilobar pneumonia.

Morphology

When visualized in infected tissue, legionellae appear as short coccobacilli to medium-length rods.

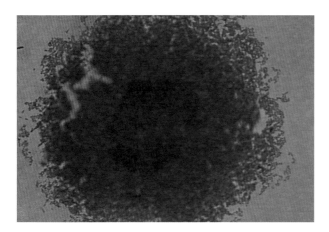

Figure 339 *Haemophilus ducreyi* High-power microscopy of colony smear, highlighting tightly bound bacilli and streaming from colony periphery. This colony morphology is the *in vitro* counterpart to the 'school of fish' seen in direct imprints of genital lesions

Figure 340 *Haemophilus ducreyi* Painful, shallow, sharply demarked ulcer with uneven edges and a grayish exudate in uncircumcised patient. Secondary lesion on glans is acquired by contiguity with ulcer on foreskin and is referred to as a 'kissing lesion'

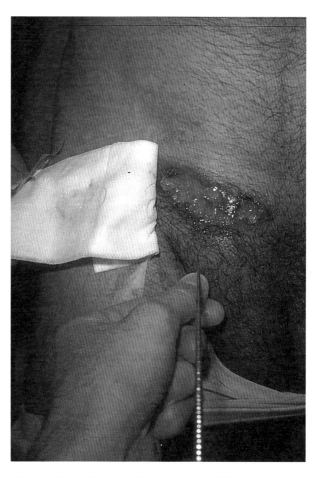

Figure 342 *Haemophilus ducreyi* Ulceration with necrotic base in right pubic area. Ulcer started 6 weeks previously as a 1-cm lesion, which progressed over a 6-week period to a 3 x 6 cm painful ulcer, which was surgically drained. This lesion, which involves the inguinal lymph nodes, is referred to as a chancroidal bubo

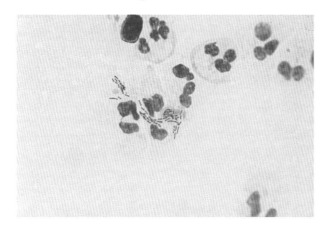

Figure 341 *Haemophilus ducreyi* 'School of fish' or railroad track aggregates of bacilli coursing between inflammatory cells in direct imprint of penile ulcer

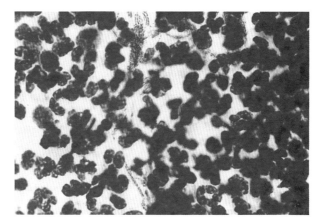

Figure 343 *Haemophilus ducreyi* Clusters of coccobacilli and bacilli in parallel rows in smear of purulent aspirate from surgical drainage of incised bubo

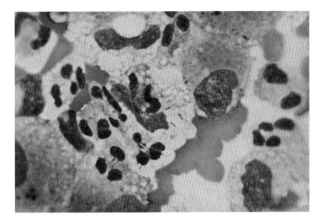

Figure 344 *Legionella pneumophila* Giemsa-stained smear of bronchoalveolar lavage showing pale blue bacillary forms within pulmonary macrophage

Figure 347 *Legionella pneumophila* Bronchoalveolar lavage culture inoculated to BCYE agar showing grayish-white, smooth to rough colonies after 72-h incubation at 35°C under 5% CO_2. No growth occurred on 5% sheep blood agar inoculated simultaneously

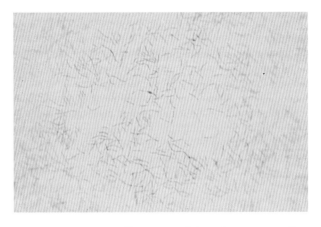

Figure 345 *Legionella pneumophila* Gram stain after growth on BCYE agar showing pale-staining, slender, slightly curved, Gram-negative bacilli

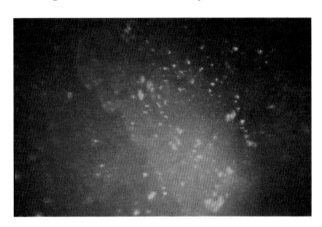

Figure 348 *Legionella pneumophila* Yellow–green immunofluorescent bacilli and coccobacilli visualized in sputum smear after staining with fluorescein-labelled monoclonal (or polyclonal) antibody

Figure 346 *Legionella pneumophila* Giemsa-stained smear of colonies growing on BCYE agar showing lavender-stained, slender, slightly curved rods. Giemsa stain allows for clearer viewing of *Legionella* microscopic morphology when contrasted to Gram-stained smears

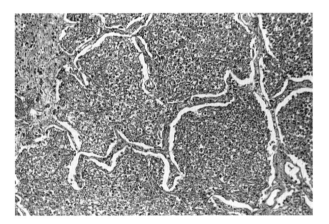

Figure 349 *Legionella pneumophila* Hematoxylin and eosin stain of lung biopsy revealing marked alveolar inflammatory response, which failed to show any microorganisms on tissue Gram stain. Note the thickened and disrupted interalveolar septa

They may occur extracellular or intracellular in mononuclear cells, in which they survive and replicate. Finding intracellular bacilli in Giemsa stains of bronchoscopy samples may suggest the presence of a *Legionella* species, which can be confirmed by immunofluorescence staining and culture. Dieterle and Warthin–Starry silver impregnation stains of histologic sections will also reveal the organism, highlighting its coccobacillary nature. Smears of growth from agar media will show a preponderance of Gram-negative, pale-staining, straight to slightly curved bacilli with rounded ends resembling pseudomonads. *L. micdadei* is weakly acid-fast (Ziehl–Neelsen method), a feature lost after cultivation on bacteriologic media. *Legionella* virulence factors include tissue-destructive enzymes. In nature, free-living amoeba (*Acanthamoeba*) serve as a host for *Legionella* growth.

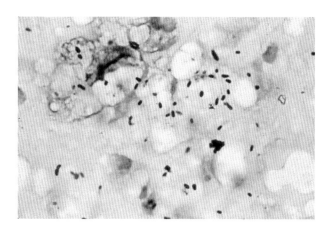

Figure 350 *Legionella pneumophila* Coccobacillary and bacillary forms in Dieterle silver stain of lung biopsy

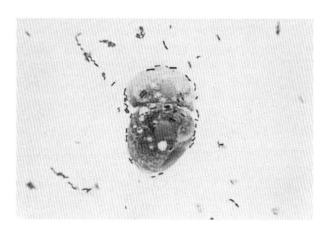

Figure 351 *Legionella pneumophila* *Acanthamoeba* trophozoite with adherent and internalized lavender-staining bacilli in Giemsa-stained smear of cocultivation in liquid media

Culture characteristics

Legionellae do not grow on routine bacteriologic media. Growth is best obtained on a selective medium, buffered charcoal-yeast extract agar (BCYE) supplemented with L-cystine, α-keto-glutarate, and ferric pyrophospate (or hemoglobin). Colonies developing on this medium are white with a speckled 'ground-glass' surface when viewed under 10–100x magnification. Confirmation of growth as a *Legionella* species can be achieved by immuno-fluorescence staining with type-specific antisera and by showing that the isolate will not grow on subculture to 5% sheep blood agar or other routinely used bacteriologic media.

MORAXELLA

Presently, the genus *Moraxella* can be separated on morphologic grounds into those species that are Gram-negative coccobacilli and those that are diplococcal. In the former are six species, of which *M. lacunata* and *M. nonliquefaciens* are rare but more common isolates, whereas the diplococcal group is represented by *M. catarrhalis*, formerly designated *Neisseria catharrhalis*, and *Branhamella catarrhalis*. These species colonize the upper respiratory tract of normal individuals. *M. catarrhalis* has emerged as a significant respiratory tract pathogen which, on culture, must be distinguished from the several other commensal and pathogenic species, including *N. meningitidis*, found in the upper respiratory tract. *M. catarrhalis* is β-lactamase-producing and a frequent

Figure 352 *Moraxella nonliquefaciens* Irregularly staining bacilli with square to slightly rounded ends, singly, in diplobacillary, and short chain arrangement admixed with pale-staining 'ghost' forms in background

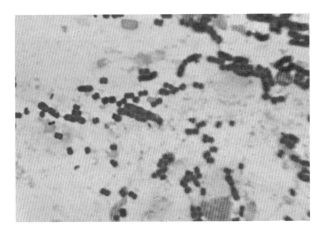

Figure 353   *Moraxella nonliquefaciens* Predominance of thick, encapsulated (red halo) bacilli and coccobacilli with square ends in direct smear of sputum from patient with chronic bronchitis. Some bacilli are Gram-variable, a characteristic of *Moraxella* species

Figure 355   *Moraxella nonliquefaciens* Mucoid, coalescing, Klebsiella-like colonies on 5% sheep blood agar from culture of sputum specimen from patient with chronic bronchitis. Isolate proved virulent for white mice by the intraperitoneal route, whereas the more typical, non-encapsulated variety was avirulent

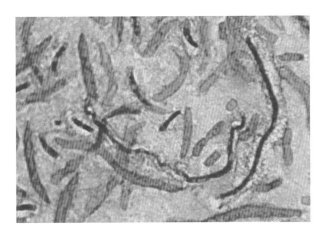

Figure 354   *Moraxella nonliquefaciens* Highly pleomorphic, irregularly stained bacillary and filament forms with rounded to pointed ends. Note pale-staining 'ghost' forms in background. Smear prepared from 5% sheep blood agar culture of highly mucoid isolate from patient with chronic bronchitis

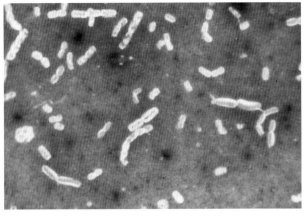

Figure 356   *Moraxella nonliquefaciens* Marked encapsulation of mucoid isolate. Smear prepared by emulsifying growth from sheep blood agar in a drop of India ink and, after drying, staining with crystal violet

cause of otitis media in children, and sinusitis and pulmonary infections in elderly individuals, especially those with chronic bronchitis and chronic obstructive pulmonary disease. Gram-stained smears of sputum samples in such instances will show a preponderance of Gram-negative diplococci in masses. Other *Moraxella* species cause primarily eye infections, including conjunctivitis and corneal ulcers (keratitis), and occasionally moraxellae may be recovered from patients with chronic bronchitis, endocarditis, septic arthritis, bacteremia, meningitis, and genital tract infections.

Morphology

Moraxella species, except *M. catarrhalis*, are characterized by their 'boxcar' morphology, appearing as square, Gram-negative bacilli and coccobacilli, with a tendency to form irregularly stained filaments, 'ghost forms', and bacilli with bulbous swellings. Bipolar staining is often seen. In patients with chronic bronchitis, smears of respiratory specimens may reveal highly encapsulated bacilli. *M. catarrhalis* is a Gram-negative, 'coffee-bean'-shaped diplococcus with

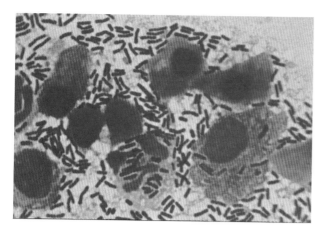

Figure 357 *Moraxella nonliquefaciens* Numerous bacilli with square ends in smear of peritoneal exudate of mouse inoculated with mucoid variant

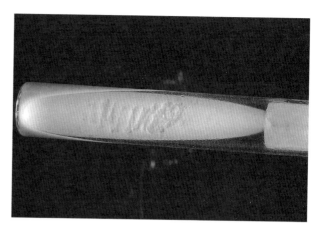

Figure 360 *Moraxella lacunata* Pitting (lacunae) colonies imbedded in Loeffler's serum medium. This species causes subacute angular conjunctivitis

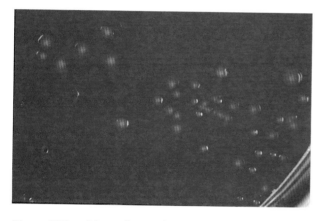

Figure 358 *Moraxella nonliquefaciens* Grayish-white smooth colonies on 5% sheep blood agar after 48-h incubation at 37°C. This colony morphotype is more typical of this species

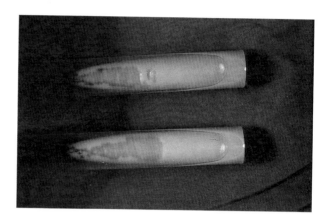

Figure 361 *Moraxella lacunata* Pitting and digestion of Loeffler's serum medium by more prodigious protease-producing strain recovered from corneal ulcer

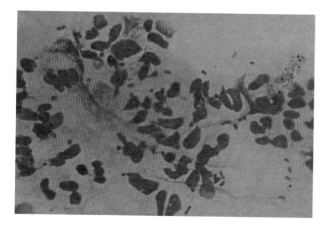

Figure 359 *Moraxella lacunata* Many diplobacilli with square ends admixed with inflammatory response in direct smear of corneal ulcer scraping

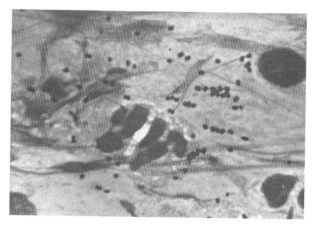

Figure 362 *Moraxella catarrhalis* Numerous 'coffee bean'-shaped diplococci singly, in pairs and small groups in direct smear of sputum from patient with chronic bronchitis. Organisms may also be observed intracellular in polymorphonuclear leukocytes

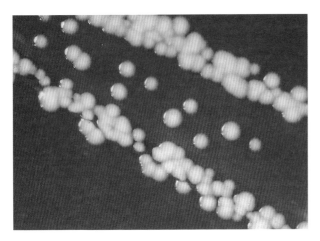

Figure 363 *Moraxella catarrhalis* Pale yellow, opaque, glistening colonies on 5% sheep blood agar after 48-h incubation. Colonies may also appear brownish and have a dry, rough consistency

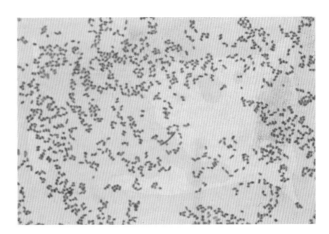

Figure 364 *Pasteurella multocida* Tiny coccobacilli and occasional bacillary forms in smear of 24-h-old 5% sheep blood agar culture

flattened, or slightly concave apposing sides, arranged in pairs, tetrads or small groupings. There is a tendency for all *Moraxella* species to retain crystal violet and therefore appear Gram-variable. Round, faintly staining, swollen cocci (involution forms) are frequently seen in culture smears of *M. catarrhalis*.

Culture characteristics

Moraxella species grow well but slowly on enriched media such as 5% sheep blood agar and chocolate agar. Growth is improved in a moist atmosphere. Colonies are raised, smooth, grayish-white with entire edges. Isolates of *M. lacunata* will either pit or digest Loeffler's serum slants. Rarely encountered, highly encapsulated *M. nonliquefaciens* may produce highly mucoid, *Klebsiella*-like colonies that coalesce in areas of contiguous growth. *M. catarrhalis* colonies are whitish-gray, becoming dry, friable, adherent and brownish.

PASTEURELLA

Members of this genus, within the family *Pasteurellaceae*, are enjoined by their small coccobacillary morphology. *Pasteurella* species are zoonotic agents colonizing the mucous membranes of the upper respiratory tract of wild and domesticated animals, including livestock, poultry and, especially, dogs and cats. It is predominantly from the latter two sources that most human infections with *Pasteurella multocida* occur, usually through direct animal bites, or by indirect contact by

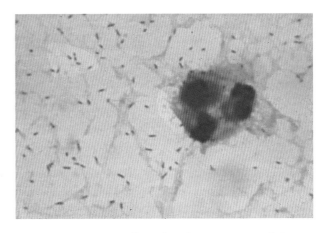

Figure 365 *Pasteurella multocida* Gram stain of abscess showing diplobacilli

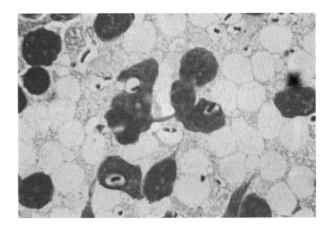

Figure 366 *Pasteurella multocida* Highly encapsulated isolate recovered from sputum of patient with chronic lung disease

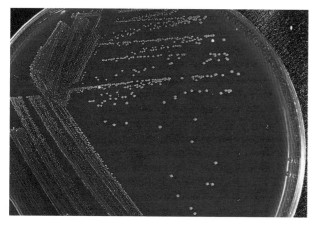

Figure 367 *Pasteurella multocida* Smooth, opaque colonies on 5% sheep blood agar

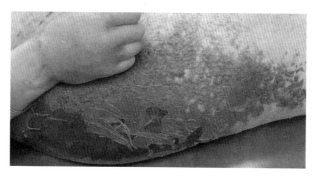

Figure 370 *Pasteurella multocida* Acute purpura fulminans (palpable purpura) which developed in an obese patient with hepatitis B cirrhosis and suspected animal (cat?) exposure. Necropsy aspirate of lesion grew pure culture of *Pasteurella multocida*. (Photo courtesy of Patrick Lento, MD)

Figure 368 *Pasteurella multocida* Mucoid, watery colonies of encapsulated strain isolated from pulmonary specimen from patient with chronic bronchitis

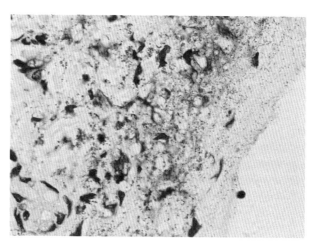

Figure 371 *Pasteurella multocida* Innumerable, small, Gram-negative coccobacilli in Gram stain of histologic section of skin biopsy of purpura fulminans lesion

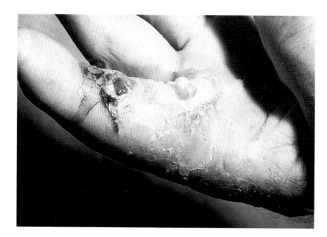

Figure 369 *Pasteurella multocida* Cat-inflicted puncture wounds of finger with purulent exudate with underlying osteomyelitis

biting or scratching of catheter tubing in use during home peritoneal dialysis. In some instances of *P. multocida* infection, an association with animal exposure cannot be established. *P. multocida* may colonize the respiratory tract of humans with chronic bronchitis and chronic obstructive pulmonary disease. In addition to localized infections surrounding a dog or cat bite, *P. multocida* may also give rise to cellulitis, abscess formation, osteomyelitis of bone underlying a penetrating bite, systemic disease including septicemia, meningitis, and rarely endocarditis. Acute epiglottitis mimicking that produced by *Haemophilus influenzae* has been reported, as have eye infections. Virulence of *P. multocida* is associated with encapsulation.

Morphology

P. multocida is a tiny coccobacillus arranged in pairs or small clusters. Longer bacillary forms may be observed. In clinical material, a capsule may be discerned. Bipolar staining is best observed with Wayson or Giemsa stains.

Culture characteristics

P. multocida grows well on 5% sheep blood and chocolate agars, forming smooth, glistening colonies within 24 h at 37°C. Isolates derived from patients with chronic respiratory tract diseases are often mucoid and watery. Colonies are non-hemolytic and growth is accompanied by a distinct musty odor, possibly related to indole production from tryptone in blood and chocolate agar. *P. multocida* does not grow on MacConkey agar.

STREPTOBACILLUS MONILIFORMIS

Streptobacillus moniliformis causes rat bite fever. The microorganism colonizes the nasopharynx of rats and other rodents. Ingestion of milk or food products contaminated by rats results in a febrile illness (Haverhill fever) characterized by fever, sore throat, polyarthritis and an erythematous skin rash. In humans, rat bite wounds introduce the microorganism into subcutaneous tissue from where it invades the lymphatics and circulation, giving rise to fever and secondary complications such as arthritis, with knees, wrists and elbows commonly affected. Other complications include endocarditis and pneumonia. *S. moniliformis* may be isoloated from blood and infected tissues using routine bacteriologic media.

Morphology

S. moniliformis is a highly pleomorphic, Gram-negative, non-motile bacterium occurring as short bacilli in chains with long filaments, often containing bulbous swellings which project from one side of the filament.

Culture characteristics

S. moniliformis grows well in liquid media enriched with blood, serum or ascites fluid in which it forms puff balls ('cotton wool balls'). Colonies on agar media are round, grayish-yellow, and glistening. 'Fried egg'-like colonies resembling those produced by *Mycoplasma* species are produced by cell wall-defective 'L'-phase variants.

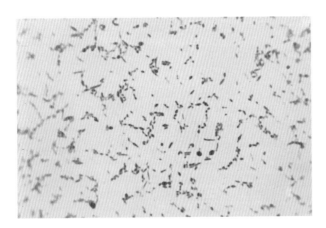

Figure 372 *Streptobacillus moniliformis* Highly pleomorphic coccobacilli and filamentous forms, many with bulbous coccoid bodies bulging into filaments and often located on one side of the filament. Some coccoid bodies are free in the background. Smear from 48-h-old culture on chocolate agar

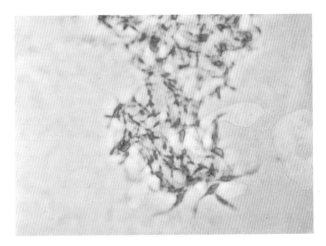

Figure 373 *Streptobacillus moniliformis* Phase-contrast-enhanced Gram stain showing marked pleomorphism and bulbous swellings projecting from one side of the filament

Figure 374 *Streptobacillus moniliformis* Grayish-white, amorphous, pinpoint colonies on chocolate agar after 72-h incubation in 5% CO_2

Figure 375 *Streptobacillus moniliformis* Characteristic 'cotton wool' puff balls in liquid media supplemented with 20% horse serum

8

Gram-positive, catalase-positive cocci

The family *Micrococcaceae* includes two genera of clinical importance, *Staphylococcus* and *Micrococcus*. The genus *Staphylococcus* contains 14 species, of which coagulase-positive *S. aureus* and the coagulase-negative species, *S. epidermidis*, *S. hemolyticus*, *S. saprophyticus*, and *S. lugdunensis* have been most frequently incriminated in human infections. Staphylococci are colonizers of the anterior nares and skin, from which they may gain access to normally sterile tissues, and through which staphylococci may be transferred from one individual to another. *S. aureus* is the predominant cause of pyogenic infections and several toxin-mediated syndromes, while *S. epidermidis* (and other coagulase-negative species) is usually associated with device-related infections, e.g. indwelling catheters, prosthesis and other foreign bodies. *S. saprophyticus* is mainly associated with spontaneous urinary tract infections in sexually active young women. The cell wall of *S. aureus* contains a unique substance, termed protein A. It is an immunologically active substance which binds the F_C fragment of immunoglobulin G. *S. aureus*, in addition to coagulase and catalase, produces a variety of exoenzymes, including hemolysins, extracellular deoxyrobinuclease, fibrinolysin, hyaluronidese, lipase, and a leukocidin. Among the protein exotoxins are those responsible for food poisoning (enterotoxin), toxic shock syndrome (toxic shock syndrome toxin 1, TSST1), and scalded skin syndrome (exfoliatin). Strains producing TSST 1 are almost universally associated with menstrual cases and for about 50% of non-menstrual cases, while strains producing enterotoxins B (47%) and C (3%) account for the remaining cases. These toxins behave as superantigens that react directly with receptors on macrophages and lymphocytes, resulting in release of inflammatory cytokines. Toxic shock syndrome (TSS) strains belong largely to bacteriophage group I. Staphylococcal scalded-skin syndrome is produced by select strains of bacteriophage group II, and originates from a focus of infection where the toxin is produced and disseminates to cause cleavage of the middle layers of the epidermidis, with resultant desquamation. Infants are most commonly affected but the syndrome occasionally occurs in adults. Methicillin resistant *S. aureus* (MRSA) and strains with intermediate resistance to vancomycin are being documented and these complicate the course of antibiotic treatment. *S. aureus* causes infections, either by direct invasion from a carrier site, usually through a break in the skin or mucosal integrity, or by production of extracellular toxins. The hallmark of staphylococcal infection is abscess formation. Primary or secondary bacteremia may result in

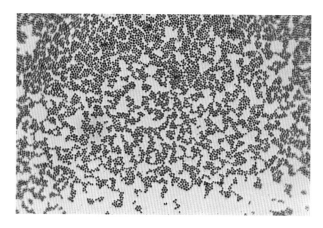

Figure 376 *Staphylococcus aureus* Gram-positive to Gram-variable cocci, singly and in 'grape-like' clusters in smear of agar culture

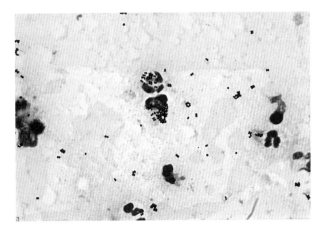

Figure 377 *Staphylococcus aureus* Cocci singly, in pairs and clusters, some intracellular, in polymorphonuclear leukocyte in smear of abscess exudate

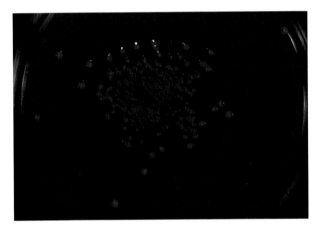

Figure 378 *Staphylococcus aureus* Yellow-pigmented, smooth, hemolytic colonies on 5% sheep blood agar after 24-h incubation

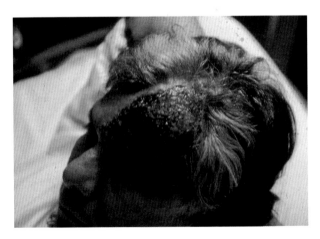

Figure 379 *Staphylococcus aureus* Pulsatile scalp abscess in patient with AIDS which developed at site of varicella-zoster reactivation

dissemination to multiple organs and bone and may lead to bacteremic pneumonia and endocarditis. Endocarditis caused by *S. aureus* and *S. epidermidis* is a special risk in intravenous drug abusers who use contaminated needles and syringes.

Morphology

Morphologically, staphylococci are characterized as forming 'grape-like' clusters of cocci because of their random cell division in three planes and because daughter cells remain attached to the parent cells. This morphologic presentation is most pronounced in smears of agar cultures, but is also present in smears of purulent exudates and broth cultures, in which cocci in pairs and short chains are also present. Individual cocci may vary in size, and some degree of decolorization may take place, rendering Gram-negative members. Some strains are encapsulated.

Culture characteristics

S. aureus is characterized by producing yellow-pigmented hemolytic colonies on 5% sheep blood agar, whereas *S. epidermidis* and *S. saprophyticus* colonies are opaque, white, and usually non-hemolytic. A peculiarity of *S. aureus* is the spontaneous production of a dwarf colony form, called a G form, which may be non-pigmented and non-hemolytic. All members of the *Micrococcaceae* family grow readily on most bacteriologic media. Mannitol salt (10% NaCl) agar is selective for *S. aureus*, which ferments mannitol, producing colonies surrounded by a yellow halo. Protein A on surface of *S. aureus* and coagulase (clumping factor) may be detected through the use of latex agglutination slide test in which latex beads coated with plasma to detect clumping factor, and antibody to detect Protein A, are aggregated by *S. aureus*.

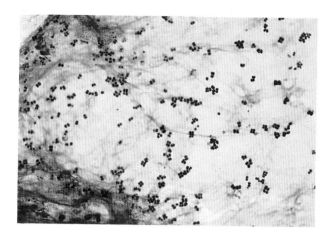

Figure 380 *Staphylococcus aureus* Innumerable cocci singly, in pairs, and clusters in smear of aspirate of scalp abscess

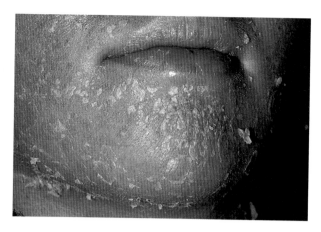

Figure 383 *Staphylococcus aureus* Desquamation of skin in patient who developed non-menstrual toxic shock syndrome subsequent to scalp abscess, induced during a vigorous scalp massage by patient's hairdresser

Figure 381 *Staphylococcus aureus* Ulcerative, postoperative circumcision infection thought to be a sexually transmitted disease, either granuloma inguinale (Donovanosis) or chancroid

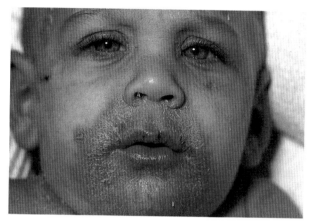

Figure 384 *Staphylococcus aureus* Perioral exfoliation of skin with denudation in child with scalded skin syndrome. The nasopharynx was the site of staphylococcal colonization. Drying and flaky desquamation follow acute episode

Figure 382 *Staphylococcus aureus* Pure culture of yellow-pigmented, hemolytic colonies developing from direct inoculation of 5% sheep blood agar with swabbings of penile lesion

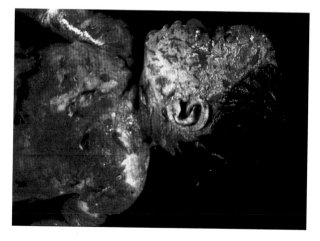

Figure 385 *Staphylococcus aureus* Fulminant presentation of scalded skin syndrome in newborn infant with staphylococcal colonization of the umbilicus. Serous fluid and electrolyte loss occur through denuded skin and often result in hypovolemia and death

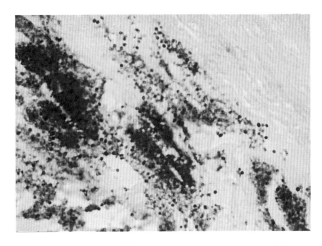

Figure 386 *Staphylococcus aureus* Gram stain of histologic section of colon biopsy from patient with enteritis, showing clusters of cocci within colonic mucosa

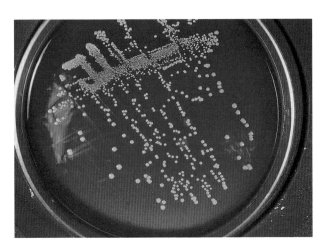

Figure 388 *Staphylococcus epidermidis* Chalk-white non-hemolytic colonies on 5% sheep blood agar. Colonies of some isolates may be sticky due to slime production by this species

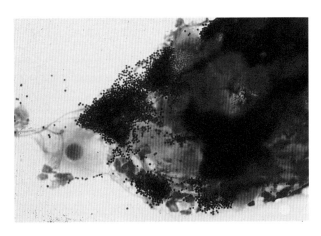

Figure 387 *Staphylococcus aureus* Innumerable Gram-positive cocci in smear of scraping of corneal ulcer

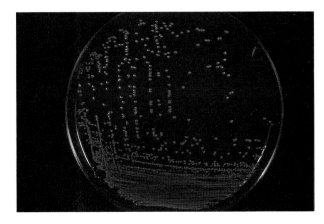

Figure 389 *Staphylococcus saprophyticus* Smooth, glossy, yellowish colonies on 5% sheep blood agar. This species is novobiocin-resistant

9

Gram-positive, catalase-negative cocci

The genus *Streptococcus*, of the family *Streptococcaceae*, is comprised of a diverse group of facultatively anaerobic to strictly anaerobic species. The genus is divided into three genera, *Streptococcus*, *Enterococcus* and *Lactococcus*. Other species of human importance include the nutritionally variant streptococci, *Abiotrophia defectiva*, *Leuconostoc*, *Pediococcus*, *Aerococcus* and *S. bovis*. Streptococci have been classified on the basis of their interaction with erythrocytes in agar media as α-hemolytic (greening due to hemoglobin conversion), β-hemolytic (clear zone of hemolysis around colonies) and γ-hemolytic (no zone of hemolysis or discoloration around colonies). Of the numerous species in the genus *Streptococcus* and *Enterococcus*, only several members account for the majority of human infections: *S. pyogenes* (Group A), *S. agalacteae* (Group B), *S. pneumoniae*, *S. bovis*, and *E. faecalis* and *E. faecium*. Viridans (α-hemolytic) streptococci are an important cause of subacute bacterial endocarditis. The Lancefield classification scheme serologically groups β-hemolytic streptococci on the basis of their carbohydrate cell wall antigens. Subtyping of hemolytic strains is based on M protein surface antigens, and greater than 90 M serotypes have been recognized. M protein confers resistance to phagocytosis and is cross-reactive with host tissues, accounting for post-streptococcal rheumatic fever and acute glomerulonephritis. Streptococcal pharyngitis may lead to rheumatic fever, a major cause of heart disease, and acute glomerulonephritis, depending on M protein serotype. Group A streptococcal M protein types 2, 49, 57, 59, 60, and 61 are associated with streptococcal pyoderma and acute glomerulonephritis, but are not associated with rheumatic fever.

Scarlet fever is caused by streptococcal pyrogenic exotoxins A, B, and C (superantigens), which are responsible for the scarlatinal rash, strawberry tongue, and desquamation of skin characterizing scarlet fever. Both throat and skin infections may lead to scarlet fever. Pyrogenic streptococcal exotoxins A, B, and C are found only in group A streptococci and serve a critical role in the pathogenesis of streptococcal toxic shock syndrome (STSS), which is characterized by hypotension, intravascular coagulation, and multi-organ failure. These toxins may also be operative in the invasion of soft tissues and skin, and in necrotizing fasciitis. Exotoxin B is an extracellular cysteine protease which cleaves fibronectin and vitronectin extracellular matrix proteins and may be important in tissue destruction. In addition to extracellular exotoxins, *S. pyogenes* releases a number of other

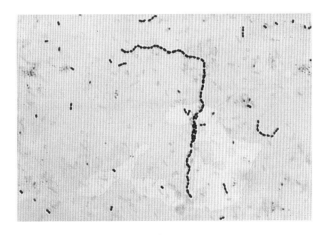

Figure 390 *Streptococcus pyogenes* Cocci singly, in pairs, and short chains in smear of broth culture

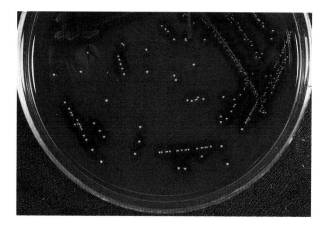

Figure 391 *Streptococcus pyogenes* Compact, smooth, β-hemolytic colonies on 5% sheep blood agar. Colony is centrally situated within a well-demarcated zone of clearing. Compact chaining within colony allows colony to be moved intact across agar surface

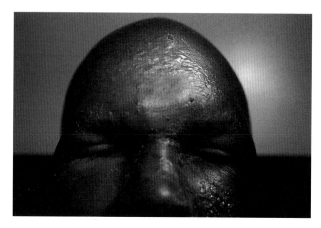

Figure 394 *Streptococcus pyogenes* Bullous cellulitis of face and scalp subsequent to manipulating pimple on left cheek. Aspirate of bullous lesion tested positive for group A streptococcal antigen by direct latex agglutination

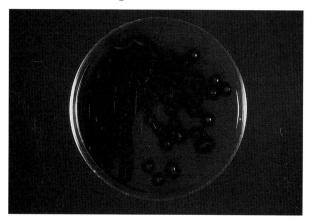

Figure 392 *Streptococcus pyogenes* Mucoid, glistening, β-hemolytic colonies of encapsulated (hyaluronic acid) strain on 5% sheep blood agar

Figure 395 *Streptococcus pyogenes* Numerous β-hemolytic colonies from direct inoculation of bullous lesion aspirate onto 5% sheep blood agar. Each colony represents a site where a drop of aspirate impinged on the agar surface subsequent to expulsion from the needle and syringe assembly

Figure 393 *Streptococcus pyogenes* Bacitracin inhibition, as indicated by zone of growth inhibition around bacitracin-impregnated disk. Test distinguishes group A strains from other hemolytic streptococcal groups

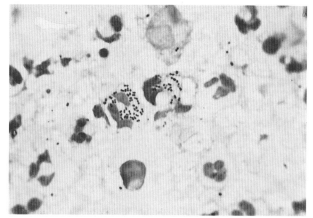

Figure 396 *Streptococcus pyogenes* Smear of bullous lesion aspirate showing cocci in pairs, singly, and in short chains

extracellular products, including streptolysin O (oxygen labile) and streptolysin S (oxygen stable). Streptolysin O is cardiotoxic and highly antigenic, rendering a brisk antibody response which can be detected. *S. pyogenes* also produces four extracellular deoxyribonucleases, and a streptokinase. The latter forms complexes with plasminogen activator and catalyzes the conversion of plasminogen to plasmin, which disgest fibrinogen and fibrin. Hyaluronidase is also produced.

Morphology

Streptococci are spherical to oval cocci which divide by the formation of an equatorial plate, resulting in chain formation, the length of which varies with species and cultural conditions; long chains predominate in broth cultures rather that in smears of agar cultures. Elongated swollen 'diphtheroid'-like forms also occur, which may confuse identification. *Leuconostoc* and *Pedicoccus* species are often elongated and coccobacillary. *Streptococcus pyogenes* produces a hyaluronic acid capsule.

Culture characteristics

With the exception of nutritionally defective streptococci, most species will grow on routine bacteriologic media, especially 5% sheep blood and chocolate agars. Nutritionally deficient streptococci (*Abiotrophia* species) require media enriched with pyridoxal (vitamin B6) or cross-streaking inocula with *Staphylococcus* as a source of vitamin B6. This results in colonies developing adjacent (satelliting) to the *Staphylococcus* streak. Colonies of Group A streptococci are opaque, smooth and compact and surrounded by a well-circumscribed zone of β-hemolysis. Encapsulated strains produce mucoid colonies surrounded by a less distinct zone of β-hemolysis. Group A streptococci are inhibited by bacitracin which serves to differentiate them from other β-hemolytic streptococci. Additionally, group A streptococci alone produce 1-pyrrolidonyl B-naphthalamine aminopeptidase which is detected by a colorimetric (PYR) disc test.

STREPTOCOCCUS AGALACTIAE (GROUP B STREPTOCOCCUS)

This streptococcal species, originally thought to be only a bovine pathogen causing mastitis, has emerged

as the single most common bacterial agent associated with neonatal bacteremia or meningitis during the first 2 months of life. Group B streptococci colonize mucous membranes of the vagina and anorectal sites of 10–25% of pregnant and non-pregnant women. Asymptomatic colonization of the genital or lower gastrointestinal tract occurs in men also and in diabetics. Newborns may acquire the microorganism *in utero* by the ascending route in the setting of prolonged rupture of membranes, or vertically by passage through the birth canal. The relative risk for colonization or invasive infection is related to the degree (inoculum size) of maternal colonization. Early-onset infection, characterized by respiratory

Figure 397   *Streptococcus agalactiae* Small cocci singly, in pairs, and in short chains in smear from 5% sheep blood agar culture

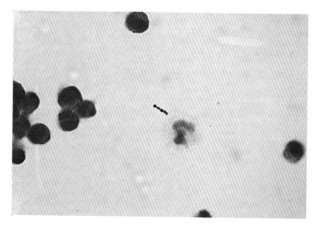

Figure 398   *Streptococcus agalactiae* Chain of cocci in direct smear of cerebrospinal fluid of infant with early-onset infection with meningitis and bacteremia

Figure 399 *Streptococcus agalactiae* Flat, opaque colonies with irregular contours surrounded by a narrow zone of β-hemolysis after 24-h growth

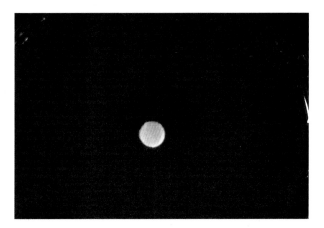

Figure 400 *Streptococcus agalactiae* CAMP test showing arrowhead zone of enhanced hemolysis of sheep erythrocytes around disk containing *Staphylococcus aureus* β-lysin placed at edge of streptococcal lawn

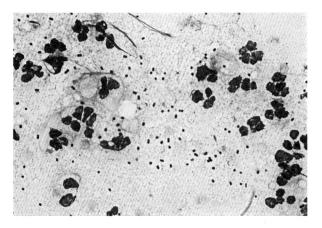

Figure 401 *Streptococcus agalactiae* Encapsulated (clear halo) cocci singly, in pairs, and short chains in smear of vaginal exudate of pregnant patient

distress, bacteremia, and meningitis, develops 20 h after birth up to 7 days post-delivery and is caused by a variety of colonizing group B serotypes. Late-onset disease most often associated with type III strains. It occurs after 7 days of birth and presents primarily as meningitis. Adult patients with group B streptococcal infections have some underlying condition, most commonly diabetes mellitus or malignancy. Infection of diabetic foot ulcers may serve as a portal of entry for bacteremic episodes. Group B streptococci may be divided into seven serotypes (Ia, Ib, II, III, IV, V, and VI), based on the presence of capsular polysaccharides. Antibiotic prophylatis has been recommended for pregnant patients with obstetric complications or who previously gave birth to a child who developed group B infection.

Morphology

Group B streptococci occur predominantly in pairs and in chain formation, especially in broth cultures. Encapsulation may be evident in direct smears of purulent exudates.

Culture characteristics

Group B streptococci grow on most bacteriologic media. On 5% sheep blood agar, colonies are gray, smooth and surrounded by a hazy, small zone of β-hemolysis which is best observed by removing surface growth and holding the Petri dish up to a light source. Group B streptococci give a positive CAMP test (named after authors Christie-Atkins-Munch-Petersen). CAMP factor is a diffusible extracellular protein that enhances hemolysis of sheep erythrocytes by staphylococcal β-lysin as evidenced by formation of an arrowhead pattern of hemolysis at the intersection of the two components. Group B streptococci are resistant to bacitracin.

STREPTOCOCCUS PNEUMONIAE

Streptococcus pneumoniae, commonly referred to as the pneumococcus, is still a leading cause of morbidity and mortality in all age groups, causing community-acquired pneumonia, otitis media, acute sinusitis, meningitis and septicemia. *S. pneumoniae* is part of the normal flora of the human nasopharynx, from where it may transgress the nasopharyngeal mucosa to gain access to the systemic circulation.

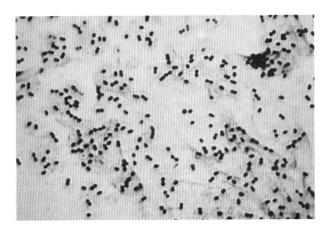

Figure 402 *Streptococcus pneumoniae* Characteristic lancet-shaped, encapsulated (halo) diplococci in smear of agar culture

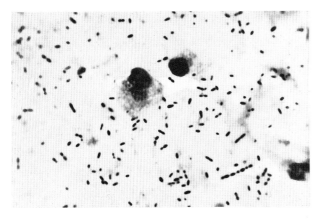

Figure 403 *Streptococcus pneumoniae* Encapsulated lancet-shaped diplococci singly, in short chains and ovoid bacilli-like forms in smear of cerebrospinal fluid of child with meningitis. The abundance of pneumococci in contrast to the paucity of inflammatory cells suggests a poor prognosis for the patient

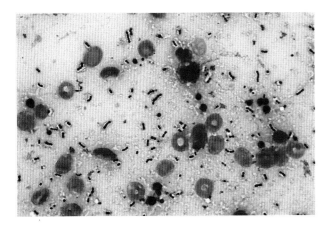

Figure 404 *Streptococcus pneumoniae* Smear of blood culture of bacteremic patient with pneumonia, showing lancet-shaped, encapsulated diplococci; confirmation as pneumococcus can be obtained by performing Quellung (capsular swelling) reaction or assaying broth for pneumococcal capsular antigen

Bacteremia leads to spread of the microorganism to the meninges, joints, peritoneum, and other sites. Meningitis may also occur by extension of an infected focus in the middle ear or paranasal sinuses. *S. pneumoniae* may also cause a purulent conjunctivitis. Individuals with compromised splenic function, e.g. as in sickle cell disease, or who are asplenic are at increased risk for spontaneous and fulminant septicemia, with potential for disseminated intravascular coagulation resembling that produced by *Neisseria meningitidis* in the course of acute meningococcemia. Virulence factors of *S. pneumoniae* include an antiphagocytic polysaccharide capsule which allows typing of pneumococci into 84 serotypes. Non-encapsulated strains are avirulent. The capsule of serotype 3 is especially large. Type-specific antibody against the capsular polysaccharide enchances phagocytosis of the bacterium and is protective against subsequent infection with the same serotype. Pneumococci spread from person to person by direct contact with infected oral secretions. Colonization and infection usually result from the acquisition of a previously unencountered serotype. Pneumococci undergo rapid autolysis in the presence of surface-active agents such as bile (sodium desoxycholate) salts and cell wall-active antibiotics. It is postulated that autolysis liberates a cytoplasmic constituent, pneumolysin, which is cytolytic for eukaryotic cells containing cholesterol in their cell membranes. Autolysis may also release mediators of inflammatory response. Polyvalent vaccines, comprised of 23 capsular polysaccharides, have been developed and are administered to individuals at increased risk for pneumococcal infections, including asplenics, the elderly, and young children. To increase immunogenicity, in pediatric populations, 7–9 valent polysaccharide vaccines conjugated to a vaccine carrier such as a bacterial toxoid are used. Penicillin-resistant strains of *S. pneumoniae* have arisen because of an alteration in penicillin binding proteins involved in cell wall synthesis.

Morphology

Streptococcus pneumoniae is an encapsulated, lancet-shaped diplococcus with its blunt ends adjacent to each other. Chains occur especially with type 3 strains, which are highly encapsulated enmeshing individual units in the capsular polysaccharide, giving rise to chain formation of varying lengths.

Figure 405 *Streptococcus pneumoniae* Smooth, small, colonies with raised periphery surrounded by greenish discoloration (α-hemolysis) on 5% sheep blood agar after 24-h incubation at 37°C and under 5–10% CO_2

Figure 408 *Streptococcus pneumoniae* Optochin (ethylhydrocupreine) inhibition of pneumococcus (type 3), indicated by large zone of growth inhibition around disk containing reagent

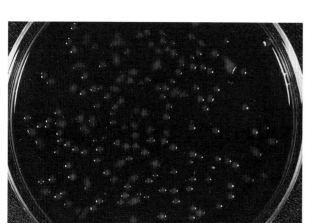

Figure 406 *Streptococcus pneumoniae* Highly mucoid, coalescing colonies of type 3 strain. Mucoid morphotype reflects pronounced encapsulation of serotype 3 strains

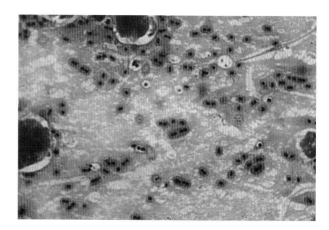

Figure 409 *Streptococcus pneumoniae* Distinctly encapsulated diplococci singly, in pairs, and short chains in smear of peritoneal exudate of mouse infected with type 3 strain

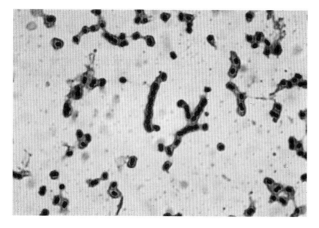

Figure 407 *Streptococcus pneumoniae* Smear of mucoid colonies showing markedly encapsulated (red halo) diplococci singly, in pairs, and in short chains

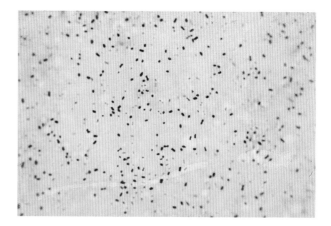

Figure 410 *Streptococcus pneumoniae* Smear of cerebrospinal fluid from child with meningitis who, prior to admission, received β-lactam antibiotic for otitis media. Note the Gram-negative forms admixed with typical staining diplococci, suggesting presence of two microbial species

Colony characteristics

Colonies developing on 5% sheep blood agar under 5–10% CO_2 incubation are smooth, with raised borders resembling a button and surrounded by a zone of green discoloration (α-hemolysis). Colonies of type 3 strains are large and mucoid, often coalescing in areas of heavy growth. Growth is inhibited by Optochin (ethylhydrocupreine), which allows for differentiation from α-hemolytic streptococci which produce similar colonies. Pneumococci are bile-soluble and cells lyse when

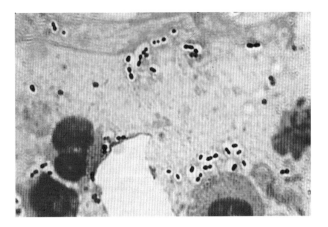

Figure 413 *Streptococcus pneumoniae* Encapsulated, lancet-shaped diplococci in smear of conjunctival scraping of patient with conjunctivitis

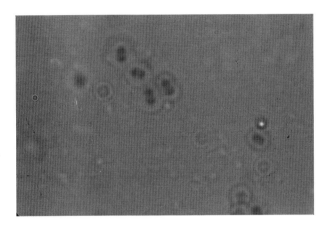

Figure 411 *Streptococcus pneumoniae* Capsular swelling (Quellung reaction) and clumping in presence of type-specific anticapsular antibody. Swollen, hydrated capsule around blue-stained pneumococcal cell is visualized microscopically as a ground-glass halo

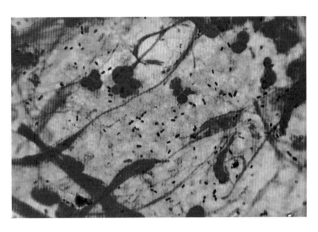

Figure 414 *Streptococcus pneumoniae* Numerous encapsulated diplococci and inflammatory cells in smear of sputum of patient with pneumonia

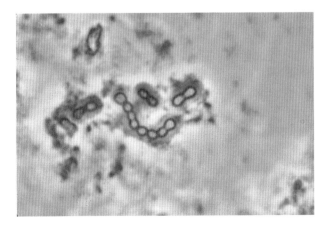

Figure 412 *Streptococcus pneumoniae* Phase contrast-enhanced microscopy of preparation of type 3 pneumococcus after interaction with type-specific antiserum, showing marked swelling (dark blue area) of capsule around chain of diplococcal cells

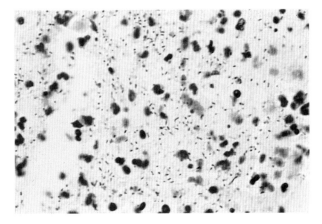

Figure 415 *Streptococcus pneumoniae* Innumerable diplococci and inflammatory cells in Gram stain of histologic section of lung section from patient with fatal pneumonia

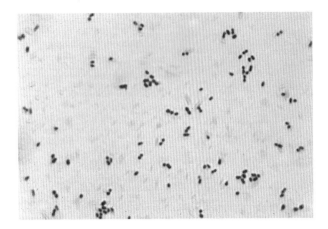

Figure 416   *Enterococcus faecalis* Ovoid cocci, singly and in pairs in smear from agar culture

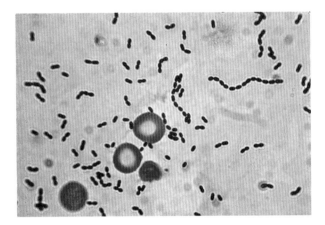

Figure 417   *Enterococcus faecalis* Ovoid cocci in pairs and short chains in smear of blood culture of patient with endocarditis. Morphologic presentation of *Enterococcus* species often resembles that of *Streptococcus pneumoniae*

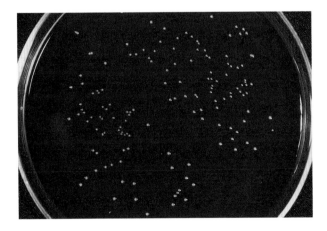

Figure 418   *Enterococcus faecalis* Small, opaque, non-hemolytic colonies on sheep blood agar after 24-h incubation. On human blood agar, β-hemolytic colonies may develop characteristics of *E. faecalis* var. *zymogenes*

placed in a 10% solution of sodium desoxycholate whereas viridans streptococci are bile-insoluble. Bile solubility results from activation of a membrane-associated amidase which splits the bond between alanine and muramic acid in the pneumococcal peptoglycan.

ENTEROCOCCUS

The genus *Enterococcus* consists of ten known species, of which *E. faecalis* and *E. faecium* are the predominant human isolates. Enterococci are common inhabitants of the gastrointestinal tract of normal individuals, with *E. faecalis* occurring in more individuals and in greater numbers/ gram of stool than *E. faecium*. Enterococci may also colonize the vagina and oral cavity. Enterococci react with group D antisera in the Lancfield typing schema. However, *Streptococcus bovis*, which also possesses the group D antigen (lipoteichoic acid) and most often produces bacteremia and endocarditis, is a non-enterococcal species and must be differentiated from enterococci because of its penicillin susceptibility and its association with carcinoma of the colon. Enterococci cause a variety of primary (endocarditis) and secondary nosocomial infections, including bacteremia, urinary tract and wound infections, and intra-abdominal infections in seriously ill and debili-tated patients. Predisposing factors also include broad use of antimicrobials, indwelling intravascular lines and urinary catheters, and surgery. Complicating nosocomial enterococcal infections is the emergence of vancomycin-resistant strains of both *E. faecalis* and *E. faecium* (VREF). The latter species is more frequently vancomycin-resistant than the former. Although enterococci produce extracellular toxins, such as cytolysin which induce tissue damage, and have the capacity to translocate across intact intestinal mucosa, they possess few definitive virulence factors. Additionally, encapsulation does not confer increased virulence.

Morphology

Enterococcus species are ovoid cocci occurring singly, in pairs, and in chains of varying length. Rare isolates of *E. faecalis* are encapsulated.

Figure 419 *Enterococcus faecalis* Mucoid morphotype on 5% sheep blood agar isolated from patient with chronic urinary tract infection. Colonies are comprised of encapsulated cells

Figure 420 *Enterococcus faecalis* Glistening mucoid colonies of encapsulated strain on chocolate agar

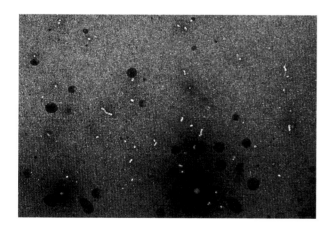

Figure 421 *Enterococcus faecalis* India ink preparation stained with crystal violet showing capsules (clear areas) around cocci, singly, in pairs and short chains

Culture characteristics

Enterococcal species grow well on most bacteriologic media, producing small, grayish-white, smooth colonies. *Enterococcus faecalis* var. *zymogenes* produces β-hemolytic colonies on human blood agar plates. Enterococci grow well in the presence of 6.5% NaCl and hydrolyze esculin in the presence of bile. They are positive for the enzyme L-pyrro-glutamyl-aminopeptidase, which hydrolyzes L-pyrrolidonyl-β-naphthylamide (PYR).

NUTRITIONALLY VARIANT STREPTOCOCCI

Abitrophia defectiva and *Granulicatella (Abiotrophia) adjacens* are part of the normal flora of the oral cavity and upper respiratory tract and are characterized on the basis of their requirement for vitamin B6 for growth on bacteriologic media. Other supporting growth factors include cysteine, thioglycollate and other sulfhydryl-containing compounds. These species have been described as satelliting streptococci, among other designations, because of their growth adjacent to a 'helper' organism producing the desired growth factor. Satellite colonies are small and either non-hemolytic or α-hemolytic. Nutritionally variant streptococci have been recovered from the blood of patients with endocarditis, including children and those with a prosthetic valve in place. Isolations have also been made from numerous other sources, including eye infections such as endophthalmitis following cataract extraction.

Morphology

In smears of positive blood cultures, a wide range of pleomorphic forms may be observed, depending on the concentration of, and nature of, the growth factor present in the medium. Morphologic presentations include oval to bulbous bacilliform cocci, singly, in short chains, and long filaments. Gram-variable staining is also evident. Smears of satelliting colonies closest to the 'helper' organism show cells of fairly uniform coccal morphology, whereas colonies selected at the perimeter of satelliting growth show bizarre morphologic entities. In some instances, smears prepared from liquid media may show extreme pleomorphism with bulbous filamentous forms resembling *Streptobacillus moniliformis*.

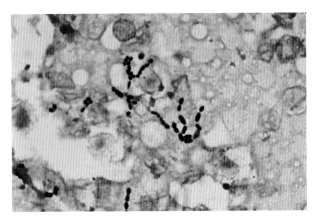

Figure 422 *Abiotrophia species* Pleomorphic, elongated, swollen cocci singly and in small clusters in smear of blood culture of patient with endocarditis

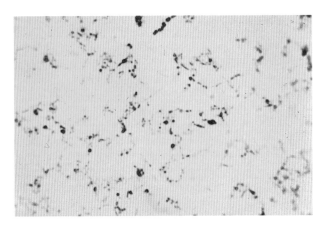

Figure 424 *Abiotrophia species* Gram-variable, highly pleomorphic cocci with bulbous swellings and bacilliform morphology in smear of satelliting colonies distal to the staphylococcal streak line

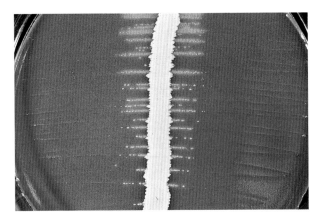

Figure 423 *Abiotrophia species* Satelliting colonies surrounded by greenish discoloration growing adjacent to staphyloccal streak on chocolate agar. Smears of colonies closest to the streak show uniform coccal morphology, whereas smears of colonies at the perimeter of satelliting growth (lowest concentration of diffusing growth factor) show marked pleomorphism

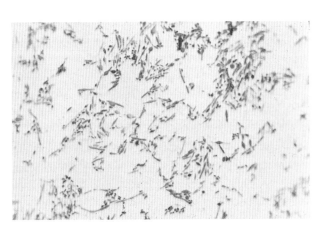

Figure 425 *Abiotrophia species* Smear of colonies on chocolate agar showing an impressive array of Gram-variable morphologic forms, ranging from disjointed filaments to shorter filaments with bulbous swellings morphologically resembling *Streptobacillus moniliformis*. Difficulties were encountered in the identification of this isolate because it did not revert to coccal morphology despite provision of growth factors. Satellitism adjacent to a staphylococcal streak proved diagnostic

Culture characteristics

Colonies on media supplemented with requisite growth factors are small, α-hemolytic or non-hemolytic. In the absence of growth factor supplementation, satelliting colonies develop around 'helper' species such as *Staphylococcus aureus*, among several, either *serendipitously* because the specimen contained other microbial species, or by design by cross-streaking inoculated plates. Growth in liquid media such as thioglycollate broth is in the form of discrete granules.

VIRIDANS STREPTOCOCCI

Viridans streptococci are members of the oral flora and upper gastrointestinal tract flora and collectively defined as those producing α-hemolytic or non-hemolytic colonies on blood agar. While greening around colonies clearly defines viridans streptococci, present classification based on hemolytic activity also includes non-hemolytic (no discoloration) species. Speciation of the individual 12 or more viridans streptococcal species is usually not undertaken, but viridans streptococci (*S. mutans*,

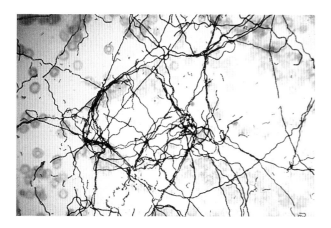

Figure 426 *Viridans streptococci* Long interwoven chains of cocci in blood culture of patient with subacute bacterial endocarditis

Figure 429 *Viridans streptococci* Clusters of cocci and chains of varying length in smear of anterior chamber fluid of eye in patient who developed an infection post-cataract surgery

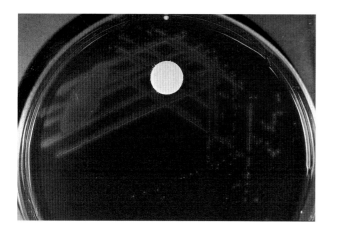

Figure 427 *Viridans streptococci* Small Optochin-resistant colonies on 5% sheep blood agar surrounded by a zone of greenish discoloration. Optochin resistance differentiates these colonies from those of *Streptococcus pneumoniae* (pneumococcus) which are inhibited by Optochin (ethylhydrocuperine)

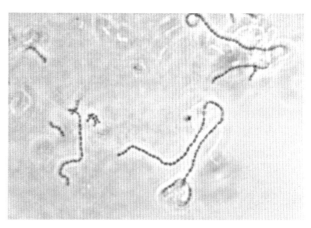

Figure 430 *Viridans streptococci* Phase-contrast-enhanced wet preparation of anterior chamber fluid showing cocci in chains, confirming streptococcal etiology of post-cataract surgical infection

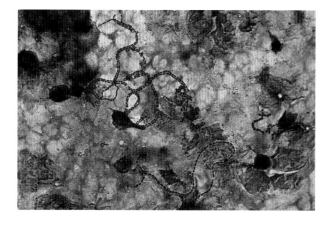

Figure 428 *Viridans streptococci* Long parallel rows of cocci in chains in Giemsa stain of bronchoalveolar lavage

S. sanguis, *S. mitis*) involved in endocarditis form exopolysaccharides (glucan, frutan), enabling them to adhere to cardiac tissue. *S. mutans* is strongly associated with smooth surface dental caries. *S. mitis* has additionally been incriminated in serious pleuropulmonary infections, endovascular infections, prosthetic hip infection, and peritonitis in the setting of small bowel perforation. Members of the *S. milleri* group (*S. intermedius*, *S. constellatus*, *S. anginosus*) have also been associated with different body sites and infections, e.g. brain and liver abscesses, genitourinary tract and gastrointestinal tract infections.

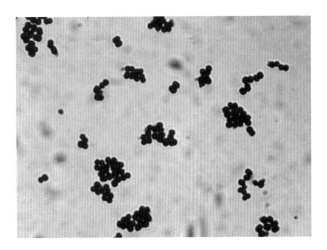

Figure 431 *Aerococcus viridans* Large cocci in pairs, clusters, and in tetrad formation in smear from 5% sheep blood agar culture

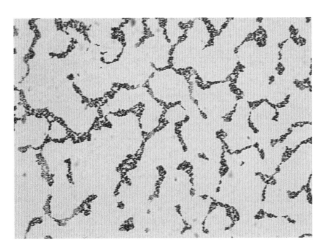

Figure 432 *Leuconostoc mesenteroides* Oval to elongated small bacilli in pairs and chains in smear from 5% sheep blood agar

AEROCOCCUS

Aerococci are Gram-positive cocci commonly found in environmental air and dust of enclosed rooms, and have been recovered from meats and vegetables. Colonization of the human respiratory tract and skin may also occur. Aerococci are catalase-negative (or weakly positive) and produce more intense greening on blood agar than colonies of viridans group streptococci. They resemble enterococci in being able to grow in the presence of 40% bile and 6.5% NaCl. In contrast to streptococci and enterococci, which produce chains of variable length, aerococci occur in tetrads and clusters resembling staphylococci. Clinically, although usually regarded as contaminants, *Aerococcus* species have been recovered from bacteremic immunosuppressed and neutropenic patients, patients with endocarditis, and from the urine of elderly patients with urinary tract infections. *Aerococcus* species are susceptibile to vancomycin.

Morphology

Aerococci are round cocci arranged in pairs or irregular clusters, with a strong tendency toward tetrad formation. Chains are not formed on agar or in broth cultures.

Culture characteristics

Aerococci grow well on most bacteriologic media. On 5% sheep blood agar, colonies are large, semi-

transparent, white or gray, and surrounded by a zone of dark greenish discoloration.

LEUCONOSTOC AND PEDIOCOCCUS

These species are catalase-negative and enjoined with other lactic acid-producing bacteria which morphologically and culturally resemble viridans streptococci and are characterized by being resistant to vancomycin. Leuconostocs are normally found among vegetable matter and in milk and other dairy products and may colonize the human oral cavity. Pediococcus species have a similar ecologic distribution as leuconostocs but may also be recovered from the human enteral tract. *Pediococcus acidilacti* and *P. pentosaceus* have been isolated from a variety of body fluids from patients with underlying diseases. Pediococci share with *Enterococcus* species esculin hydrolysis, 6.5% NaCl tolerance, and the group D antigen.

Morphology

Leuconostocs are Gram-positive, spherical to elongated rod-shaped cells, appearing morphologically between streptococci and lactobacilli. Small bacillary forms predominate on agar media and in smears of broth cultures, allowing differentiation from streptococcal species. *Leuconostoc mesenteroides* has been isolated from septicemic immunosuppressed patients, especially those with an indwelling catheter, and from cerebrospinal fluid from a previous healthy woman. *Leuconostoc* species may be differentiated from streptococci by gas production from glucose. Pediococci are Gram-positive, round cocci that characteristically form tetrads and small clusters.

Culture characteristics

Leuconostoc and *Pediococcus* colonies resemble those of viridans streptococci, being pinpoint after 24-h incubation, white to gray, and non-hemolytic. A greenish discoloration develops around colonies after 48 h or more incubation.

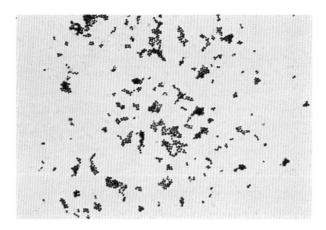

Figure 433   *Pediococcus species* Round cocci in pairs and tetrad formation in smear from 5% sheep blood agar culture

10

Gram-negative cocci

The genus *Neisseria* in the family *Neisseriaceae* is comprised of 14 genera of Gram-negative diplococci, of which *N. gonorrhoeae* and *N. menigitidis* are the two bonafide human pathogens. *N. gonorrhoeae* (gonococcus) causes gonorrhea, a highly prevalent venereal disease, while *N. menigitidis* (meningococcus) is the etiologic agent of acute meningitis and fulminant meningococcemia. The only known reservoir for *Neisseria* species are humans: the upper respiratory tract and genitourinary tract. Rectal carriage of *N. gonorrhoeae* and *N. meningitidis* also occurs. *Neisseria* species are aerobic and facultatively anaerobic and produce oxidase. *N. gonorrhoeae* is primarily a cause of urethritis in males, and cervicitis and pelvic inflammatory disease in females. In these sites, gonococci adhere to microvilli of columnar epithelium through surface pili, or hair-like filaments extending from the cell surface. Non-piliated organisms are avirulent. Pilin protein structure is antigenically different between strains. Additionally, *N. gonorrhoeae* can colonize and infect cervical, conjunctival, and rectal mucosa, leading to a purulent inflammatory response. In prepubertal females, *N. gonorrhoeae* can infect the cuboidal vaginal epithelium, causing vaginitis. Strains with specific types of outer membrane protein II produce opaque colonies, whereas strains lacking these proteins produce transparent colonies. These proteins also enhance gonococcal attachment to buccal epithelial cells and aid intergonococcal adherence, resulting in clumps of organisms termed 'infectious units'. Outer membrane proteins may also enhance dissemination from colonizing sites, especially in the oral cavity and mediate resistance to serum bactericidal activity. Mucosal pathogens such as *N. gonorrhoeae* and *N. meningitidis* also produce immunoglobulin A protease. *N. gonorrhoeae* is spread from person to person through contact with infected secretions, usually through the sexual route from asymptomatic or symptomatic individuals. Oral–oral and oral–genital transmission may also occur. Colonization and infection of the oropharynx may lead to dissemination to joints (usually one) rendering septic arthritis, to the pericardium, or aortic valve, causing acute endocarditis, and to the skin, giving petechial lesions, usually on the extremities. Rarely, fulminant gonococcal septicemia with disseminated intravascular coagulation occurs. Conjunctivial epithelium is highly susceptible to gonococcal invasion, rendering a purulent conjunctivitis in newborns (ophthalmia neonatorum) and in other age groups as well.

Morphology

Neisseria species are diplococci with flattened or slightly concave adjacent sides and may occur in pairs, tetrads, or small clusters, with a tendency to retain crystal violet during Gram staining. As cultures age, the diplococcal forms undergo autolysis, giving rise to large swollen bodies (involution forms) which stain faintly Gram-negative. *N. meningitidis* is encapsulated.

Culture characteristics

Neisseria gonorrhoeae and *N. meningitidis* may be regarded as fastidious microorganisms requiring enriched media such as 5% sheep blood and chocolate agars for growth and incubation at 35–37°C under 5% CO_2. Both species grow well on Thayer–Martin medium which is an enriched medium and which contains several antimicrobials

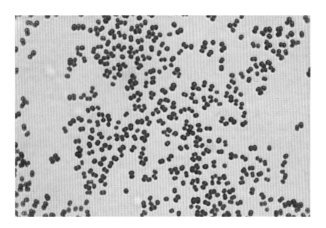

Figure 434 *Neisseria gonorrhoeae* Typical diplococci with flattened inner surface resembling a 'coffee bean' in smear of agar culture

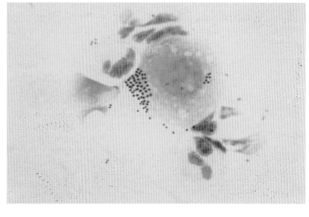

Figure 437 *Neisseria gonorrhoeae* Large cluster ('infectious unit') of multiplying gonococci derived from rupture of phagocytic cell and which have a special predeliction for urethral and conjunctival epithelial cells. Such clusters, seen in direct smears of purulent exudate, may also be surrounded by remnants of the phagocytic cell

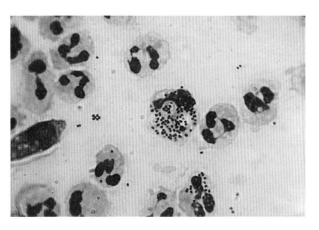

Figure 435 *Neisseria gonorrhoeae* Intracellular and extracellular diplococci in smear of urethral discharge of patient with gonorrhea. Similar morphology will also be seen in smears of purulent conjunctival exudates

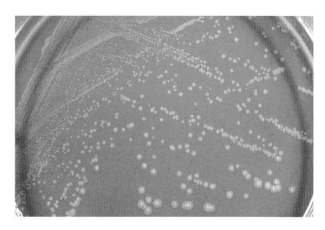

Figure 438 *Neisseria gonorrhoeae* Chocolate agar culture showing large, non-piliated colony morphotype with irregular edge and small, glistening, piliated colony morphotype

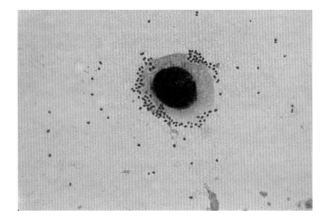

Figure 436 *Neisseria gonorrhoeae* Conjunctival scraping showing numerous diplococci adherent to, and invading, conjunctival epithelial cell. Many free diplococci are also visible

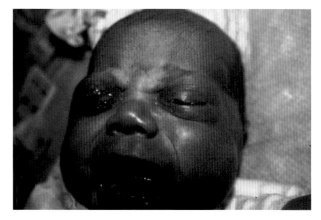

Figure 439 *Neisseria gonorrhoeae* Ophthalmia neonatorum in an infant caused by a penicillinase-producing isolate. Note swollen eyelids, and copious eye discharge (courtesy Harold Raucher, MD)

to suppress the growth of commensal micro-organisms, especially when attempting to isolate *N. gonorrhoeae* from sites, e.g. genital, rectal, pharyngeal, containing other microbial species. Freshly isolated colonies of *N. gonorrhoeae* are piliated and are small, smooth, and glistening, whereas non-piliated colonies, which may develop spontaneously, or after subculture, are larger, rough with irregular borders. *N. meningitidis*, although piliated, does not show colony transition based on presence or absence of pili. Colonies are smooth, entire, gray–brown, and of

a watery consistency reflective of encapsulation. *N. meningitidis*, in contrast to *N. gonorrhoeae*, grows more luxuriantly on 5% sheep blood agar. *Neisseria* species are oxidase- and catalase-positive.

N. meningitidis colonizes the upper respiratory tract from where it can disseminate to cause acute meningitis or fulminant bacteremia (meningococcemia), characteristically associated with a petechial or purpuric rash and usually accompanied by intravascular coagulation (blood vessels blocked by fibrin thrombi, leukocytes, and bacteria).

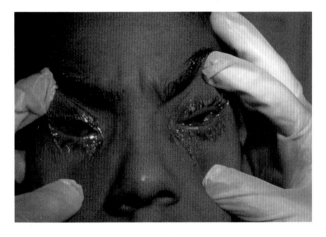

Figure 440 *Neisseria gonorrhoeae* Bilateral conjunctivitis in a 19-year-old patient who also presented with genital infection

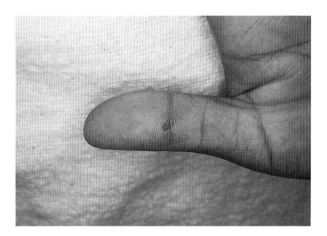

Figure 442 *Neisseria gonorrhoeae* Skin lesion on thumb of patient with disseminated gonococcal infection. Lesions are few in number in contrast to disseminated meningococcal infection and usually occur on extremities. They may begin as small, tender papules or petechiae that become pustular and hemorrhagic

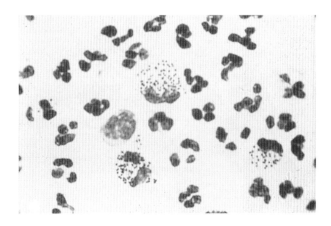

Figure 441 *Neisseria gonorrhoeae* Intracellular diplococci in direct smear of conjunctival exudate from patient with bilateral conjunctivitis. Cultural confirmation must be undertaken to differentiate gonococcal conjunctivitis from primary meningococcal conjunctivitis, as the latter microorganism has a greater potential for systemic spread

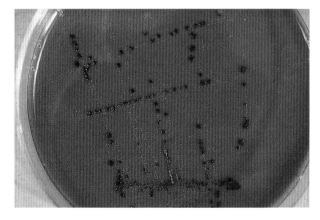

Figure 443 *Neisseria gonorrhoeae* Dark purple to black colonies indicative of oxidase production after flooding colonies with indicator dye tetramethyl-*p*-phenylenediamine dihydrochloride. All *Neisseria* species are oxidase-positive

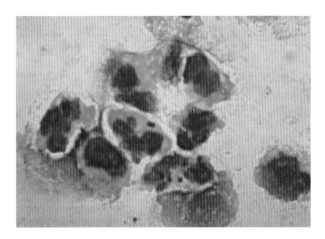

Figure 444   *Neisseria meningitidis*   Phagocytized diplococci in direct smear of purulent cerebrospinal fluid of patient with acute meningitis

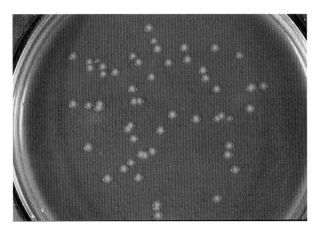

Figure 445   *Neisseria meningitidis*   Gray, smooth, translucent colonies with even periphery on chocolate agar after 48-h incubation at 37°C under 5% CO_2

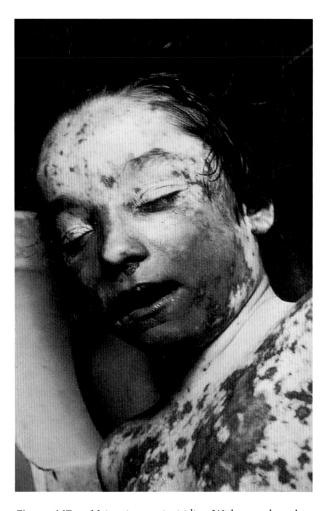

Figure 447   *Neisseria meningitidis*   Widespread ecchymotic purpuric skin lesions (petechiae) in complement-deficient child with fulminant meningococcemia and Waterhouse–Friderichsen syndrome

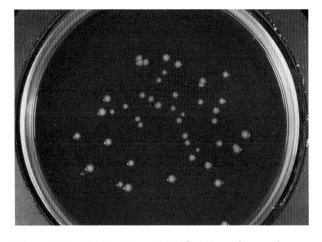

Figure 446   *Neisseria meningitidis*   Smooth, translucent colonies on 5% sheep blood agar after 48-h incubation

N. meningitidis may be divided into 13 serogroups (A, B, C, D, 29E, H, I, K, L, W135, X, Y, and Z), based on the presence of an anionic polysaccharide capsule. Serogroups A, B, C, W135 and Y cause the majority of human infections. Invasive meningococci are encapsulated. Aerosolization of respiratory secretions is the major route of person-to-person transmission. Colonization of the human oropharyngeal mucosa is through attachment by surface pili (fimbriae) of the microorganism to specific receptors on the oropharyngeal mucosa. Outcome of contact with *N. meningitidis* depends on strain virulence, status of host immunity, e.g. complement deficiency, especially of late components (C6, C7, C8), immunoglobulin M

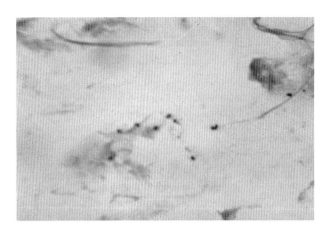

Figure 448 *Neisseria meningitidis* Diplococci in Gram stain of scraping of petechial lesion of patient with acute meningitis, rendering rapid diagnosis of meningococcal infection

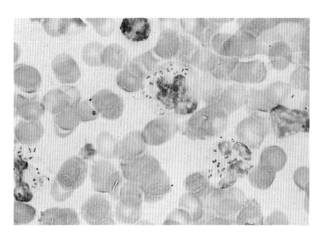

Figure 449 *Neisseria meningitidis* Phagocytized diplococci in Wright stain of peripheral blood smear of child with overwhelming fatal meningococcemia

deficiency, and preceding respiratory infection which may facilitate invasion. Serum bactericidal antibody directed against capsular or non-capsular antigens is protective. Invasive meningococci can form metastatic lesions in the skin, eyes, joints, heart, adrenals, or meninges. Bilateral adrenal hemorrhage (Waterhouse–Friderichsen syndrome) occurs in 3–4% of individuals with systemic meningococcal infection. Pneumonia, urethritis proctitis, and primary conjunctivitis are other manifestations of meningococcal infection. Chronic meningococcemia is an unusual manifestation which may resolve spontaneously or continue for months with episodes of intermittent fever.

11

Curved and spiral species

CAMPYLOBACTERIACEAE

Based on RNA sequence studies, the family *Campylobacteriaceae* is comprised of 16 species and six subspecies of *Campylobacter*, six species of *Helicobacter* and three of *Arcobacter* species. All are enjoined by being microaerophilic, Gram-negative, motile, curved rods. *C. jejuni*, *C. coli*, and *C. fetus* are the most common *Campylobacter* isolates associated with diarrheal illness and septicemia, while *H. pylori* has been solidly linked to peptic ulcer disease and gastric carcinoma. The primary reservoirs for campylobacters are animals (cattle, pigs), especially poultry, from which human infections occur through contaminated food or water, or by ingestion of unpasteurized milk. Outbreaks have occurred in association with consumption of chicken or raw milk as well as from contact with infected cats and dogs. Drinking untreated water from rivers and streams in rural areas where animal contamination occurs is also a major route of acquisition.

Campylobacter species

Campylobacter infections are global in scope and, in many geographic locales, including the United States, exceed diarrheal illness caused by *Salmonella*. Infants are particularly targeted, with young adults the second largest group affected. Infections can range from asymptomatic carriage to bloody diarrhea with sepsis and death. Diarrhea may be severe with eight to 15 bowel movements daily and is indistinguishable from that caused by *Salmonella* or *Shigella* species. Fecal leukocytes are frequently present, as is gross blood. Extraintestinal infections include bacteremia, septic thrombophlebitis, and meningitis, and frequently occur in individuals with predisposing conditions, e.g. alcoholism, immunosuppression, extremes of age. *C. fetus*, which is resistant to serum bactericial activity, often presents as bacteremia, although the majority of blood isolates are *C. jejuni*. Post-infection sequelae secondary to gastrointestinal involvement include Guillain–Barre syndrome, reactive arthritis, and hemolytic uremic syndrome. Virulence factors include motility, which aids invasion of intestinal epithelium leading to an acute inflammatory enteritis of the small intestine and colon. Shedding of organisms can last up to 3 weeks.

Morphology

Campylobacter species are characteristically curved to spiral bacilli which display a darting motility in wet preparations of cultures or stool specimens examined under phase-contrast or dark-field microscopy. In Gram-stained smears, S-shaped 'seagull wing' forms may be visualized, along with curved and spiral forms. Smears of cultures several days old may show only spherical forms.

Culture characteristics

Campylobacters grow best in an atmosphere containing 5% oxygen, 10% carbon dioxide and 85% nitrogen. Selective culture media containing antibiotics (vancomycin, trimethoprim, polymyxinB, cephalothin, and amphotericin B) and 10% sheep blood (CAMPY-BAP) are commonly used, although other formulations are also available. *C. jejuni* and *C. coli* are thermo-tolerant and grow at 42°C whereas 37°C is required for *C. fetus*. On these media, colonies vary in color from gray to pinkish, and, on

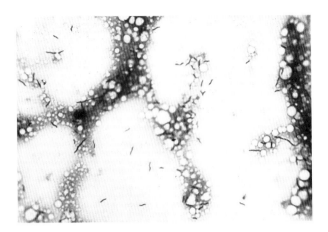

Figure 450 *Campylobacter jejuni* Slightly curved, spiral rods in direct smear of blood culture from immunosuppressed patient with enterocolitis

Figure 453 *Campylobacter jejuni* Innumerable, smooth, opaque, slightly spreading colonies on CAMPY-Blood agar from stool of medical student with bloody enterocolitis who, while on a hiking trip, camped adjacent to a stream and drank water from the stream

Figure 451 *Campylobacter jejuni* Bloody dysenteric-like stool from medical student who developed diarrhea after drinking water from a stream while camping outdoors

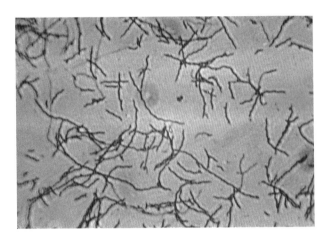

Figure 454 *Campylobacter fetus* Spiral, slightly curved rods in direct smear of blood culture of immunosuppressed patient. Direct visualization of wet preparation of blood culture broth was remarkable for rapidly motile bacilli darting across the microscopic field

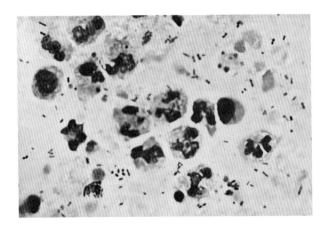

Figure 452 *Campylobacter jejuni* Small, slightly curved bacilli admixed with purulent exudate in direct smear of stool from medical student with bloody enterocolitis

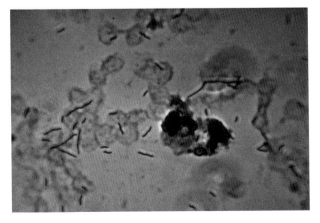

Figure 455 *Campylobacter fetus* Phase-contrast-enhanced Gram stain of 48-h-old blood culture showing predominance of curved spiral forms

media of high moisture content, may be flat, grayish, watery and spreading, with an irregular shape. *Campylobacter* species are oxidase- and catalase-positive.

Helicobacter

The genus is comprised of 17 species of which *H. pylori*, a urease-producing species, is the most prominent human pathogen linked to type B antral gastritis, peptic ulcers, and gastric carcinoma. A linkage to coronary heart disease has also been documented. Other non-urease-producing members, *H. cinaedi* and *H. fennelliae*, formerly classified as *Campylobacter* species, are enteric pathogens. *Helicobacter pylori* is distributed world-wide and colonizes the gastric antrum beneath the mucus layer, which it penetrates through a rotary motion facilitated by multiple polar flagella. Additionally, *H. pylori* possesses surface adhesins (pili, hemagglutinins) which aid adherence to gastric epithelia facilitated by specific blood group O type receptors. Urease production is a major virulence factor, enhancing colonization by degrading tissue urea to carbon dioxide and ammonia which neutralizes gastric acidity. Two *H. pylori* biotypes, I and II, have been described. The latter produces an intracytoplasmic vacuole-inducing cytotoxin which experimentally induces gastric ulceration in mouse models. Serum neutralizing antibody is induced by the vacuolating cytotoxin. Type I strains are, however, more virulent and closely associated with gastric ulcers and carcinomas. Although 40–50% of gastric malignancies are linked to *H. pylori* infection, less than 1% of infected individuals develop this complication, which takes decades to ensue. *H. pylori* has been recovered from feces of adults and children. Transmission has also occurred among patients via endoscopy (fiberoptic gastroduodenoscopy). Other putative modes of transmission include water and water-contaminated foods.

Morphology

Helicobacter are slightly curved, spiral bacilli which are motile by five to seven terminal flagella. In older cultures, coccoid forms predominate and these maintain infectivity but are not cultivable on subculture. In direct smears of gastric biopsies, the microorganism is arranged linearly within the mucus layer. In addition to the Gram stain, *H. pylori* can be detected with silver and Giemsa stains.

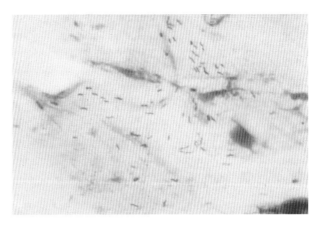

Figure 456 *Helicobacter pylori* Touch imprint of gastric biopsy showing numerous curved bacilli singly, in parallel groupings, and linearly aligned in mucus layer

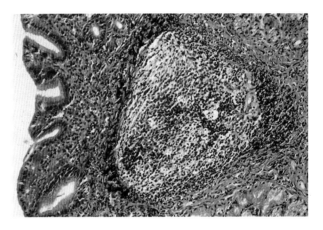

Figure 457 *Helicobacter pylori* Hematoxylin and eosin stain of antral biopsy in a 13-year-old child with gastric lymphonodular hyperplasia, showing prominent lymphoid follicle surrounded by marked inflammatory response. The occurrence of lymphoid follicles is suggestive of local stimulation of humoral immune response to the microorganism

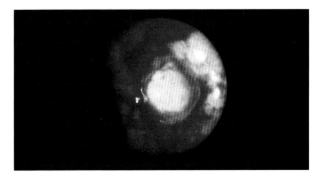

Figure 458 *Helicobacter pylori* Endoscopy of gastric antrum of 11-year-old patient showing multiple ulcerative lesions (white areas) with significant amount of hemorrhage

Colony characteristics

Helicobacter grow best under microaerophilic conditions at 37°C on enriched chocolate agar with increased humidity. Colonies are small, translucent and circular and take 4–7 days to develop. A faint zone of hemolysis may be observed around colonies on 5% sheep blood agar. Gastric biopsy specimens may be inoculated directly to Christensen's urea agar which will turn bright pink within minutes indicative of urease production and, hence, the presence of *H. pylori*.

Spirillum minus

Once regarded as a spirochete, *S. minus*, the cause of rat-bite fever (sodoku), is a tightly coiled, Gram-negative, motile microorganism. Motility is darting and *Vibrio*-like through the action of polar flagella. Rat-bite fever is an acute febrile illness with chills that results 1–4 weeks after a rat bite. *S. minus* is a natural parasite of rats, which are asymptomatic carriers of the microorganism. Subsequent to a rat bite, an ulcerative lesion may be present at the inoculation site, and local lymphadenopathy and lymphangitis develop with systemic disease. Fever may decline, then recur in episodes for months to years. Diagnosis is achieved by direct visualization of spirocletes in blood or tissue (lymph nodes) samples using Giemsa stain or under dark-field microscopy. To date, *S. minus* has not been cultivated on artificial media. *S. minus* rat-bite fever is rarely encountered in the United States, and most infections are documented in Asia.

SPIROCHETES

Borrelia, *Leptospira*, and *Treponema* species are slender, helical or spiral bacteria within the order Spirochetales, which is subdivided into two families: *Spirochetaceae*, which includes the genera *Borrelia* and *Treponema*, and *Leptospiraceae* which includes the single genus *Leptospira*. The ultrastructural morphology of spirochetes is unique: the cytoplasm of the cell is surrounded by a cytoplasmic membrane and a peptidoglycan layer, which helps to maintain cell rigidity and shape. In *Treponema* species, axial filaments are present in the bacterial cytoplasm, which is covered by a triple-layered, external outer envelope. The axial filaments are located between the outer envelope and the cell wall–membrane

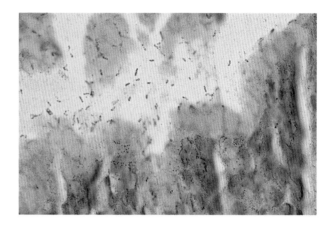

Figure 459 *Helicobacter pylori* Dieterle silver stain of antral biopsy showing numerous slightly curved, spiral rods along disrupted gastric mucosa

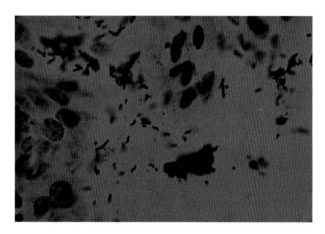

Figure 460 *Helicobacter pylori* Higher magnification of Dieterle-stained gastric biopsy, depicting slightly curved, short bacilli and 'seagull forms'

complex. Individual cells are curled helically around one or more axial filaments. Outside the outer membrane is a mucoid layer. Spirochetes are motile by contraction and extension of the microorganism's body about its axial filaments. Virulent treponemes have three axial filaments, borreliae from seven to 30, and leptospires only one. Axial filaments are anchored to pores at the ends of the spirochetes. Although treponemes are distantly related to Gram-negative bacteria, only *Borrelia* species are stained by the Gram method. *Treponema* and *Leptospira* species are more readily visualized by dark-field or phase-contrast microscopy. For all three spirochetal genera, human infections have overlapping attributes and species-specific features. *Treponema*, *Borrelia* and *Leptospira* enter through the skin or mucus membranes where primary lesions occur. Spirochetemia ensues with focalization to target organs, followed by a latent period with subsequent secondary and tertiary disease. *Treponema pallidum* subspecies *pallidum* causes syphilis; *T. pallidum* subspecies *endemicum* causes bejel (endemic syphlis); *T. pallidum* subspecies *pertenue* causes yaws; *T. carateum* causes pinta. *Borrelia recurrentis*, *B. hermsii*, and *B. duttonii* are associated with louse-borne epidemic relapsing fever (*B. recurrentis*) and endemic tick-borne relapsing fever; *B. burgdorferi* with Lyme disease. *Leptospira* species cause leptospirosis.

Borrelia burgdorferi

B. burgdorferi is a long, corkscrew-shaped spirochete with seven to 11 axial filaments between the outer envelope and the cytoplasmic membrane. *B. burgdorferi*, the causative agent of Lyme disease, grows within and is transmitted by *Ixodes* ticks: *Ixodes scapularis* along the eastern coast of the United States and *Ixodes pacificus* in the northwest. Primary reservoirs for the spirochete are the white-footed mouse (*Peromyscus leucopus*) and other small rodents. Tick feeding patterns are germane to the distribution of the disease. *Ixodes* ticks have three feeding stages over 2 years: larval, nymphal, and adult. Larval ticks become infected after birth by feeding during late summer and early fall on an infected animal, usually a small rodent, that harbors *B. burgdorferi* asymptomatically. Taking a blood meal, larval forms then molt the following spring into eight-legged nymphs, which feed on larger animals such as racoons and squirrels, as well as on small rodents and humans, passing *B. burgdorferi* onto their hosts. Nymphs are tiny, resembling a pepper grain, and often go unnoticed. After this second blood meal, the infected nymphs molt in late summer into adults, which then feed on white-tailed deer and humans. To transmit *B. burgdorferi* to humans, the adult tick must remain attached for 48–72 h while engorging itself with blood. Its larger size, however, favors it being noticed and removed prior to *Borrelia* transmission. Subsequent to transmission, lyme *Borrelia* induce a variety of clinical manifestations including (stage 1) skin lesions (erythema chronicum migrans), neurologic and cardiac abnormalities (stage 2) and persistent arthritis (stage 3).

Borrelia recurrentis

Louse-borne epidemic relapsing fever is caused by *B. recurrentis*, while endemic, sporadic, tick-borne

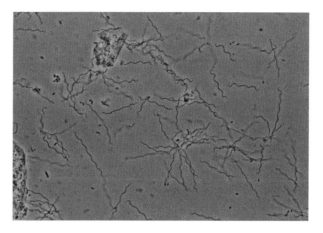

Figure 461   *Borrelia   burgdorferi*   Phase-contrast microscopy of liquid culture showing loosely coiled spirochetes of varying lengths

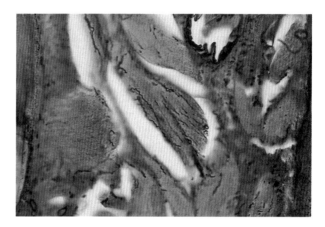

Figure 462   *Borrelia burgdorferi* Numerous spirochetes linearly arranged in hind gut of *Ixodes scapularis* tick removed from skin, sectioned, and silver-stained by the Warthin–Starry method (courtesy of Robert Phelps, MD)

Figure 463   *Borrelia burgdorferi* *Ixodes scapularis* adult tick with hard dorsal plate or scutum and clearly visible mouth part. Adult tick has eight legs (right leg out of focus)

Figure 464 *Borrelia burgdorferi* High-power magnification of *Ixodes scapularis* showing serrated mouth part

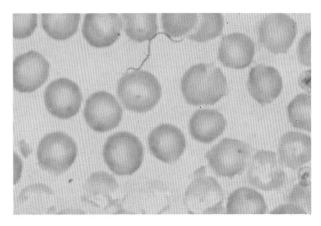

Figure 467 *Borrelia recurrentis* Curved, slender spiral form in Giemsa stain of peripheral blood smear

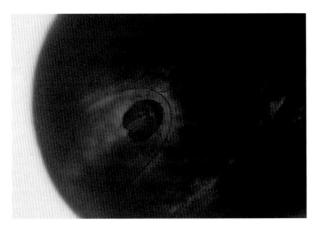

Figure 465 *Borrelia burgdorferi* Dorsal plate and anus of *Ixodes scapularis* tick surrounded by semi-circular anal groove

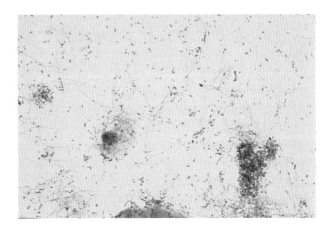

Figure 468 *Borrelia vincenti* Numerous loosely coiled, slender spirochetes of varied lengths in direct smear of vaginal exudate from patient with symptomatic vaginitis

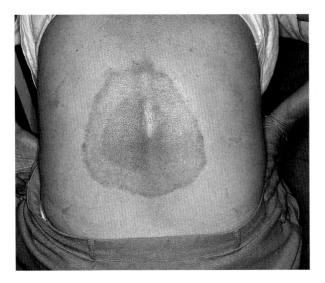

Figure 466 *Borrelia burgdorferi* Erythemia chronicum migrans lesion on back of patient which has developed subsequent to tick bite. (Courtesy of Gary P. Wormser, MD)

relapsing fever is caused by *B. hermsii* and *B. duttonii*. In the former, transmission takes place by squashing infected lice with a fingernail, while the latter is transmitted by the bite of an infected tick. As lice do not transmit *B. recurrentis* to their offspring, they must continuously feed on infected hosts to maintain the infectivity cycle among the louse population. In contrast to *B. recurrentis*, *B. hermsii* and *B. duttonii* are passed on transovarially to newborn ticks, thereby confining the infection within the vector (*Ornithodorus*) population. Ticks are soft-bodied, feed rapidly (10–20 min), are primarily nocturnal, and, when biting, cause no discomfort and hence go unnoticed. Tick-borne relapsing fever is endemic in Colorado and Arizona, especially along the Grand Canyon. Rodents can serve as reservoirs for the infectious agent. Louse-borne relapsing fever has vanished from the United States. The high incidence of louse-borne relapsing fever occurring in endemic areas such as Latin America is enhanced by crowding and unhygienic conditions. Subsequent to the bite of an infected louse or tick, spirochetes enter the blood and lymphatic circulation, multiply to high concentrations, and fever begins. Those untreated patients who survive develop relapsing episodes of fever and chills related to the change in the immuno-dominant protein (variable major proteins) in the *Borrelia* outer membrane present at the time of infection. Host antibody response to this immuno-dominant protein results in complement-mediated lysis of the majority of borreliae. Some borreliae persist because they express a new protein, not recognized by the circulating antibody, multiply to enormous numbers (10^6–10^8) and the patient then relapses. Genes accounting for antigenic shift are carried on plasmids. Diagnosis of *Borrelia* infection can be achieved by demonstrating spirochetes in Giemsa or Wright-stained smears of peripheral blood.

Morphology

Borrelia species are organisms whose spirals are coarse, shallow and irregular, and whose refractive index is similar to that of bacteria and hence are stained Gram-negative by the Gram method. Diagnosis is usually achieved by demonstrating borreliae in stained blood films, rather than by cultivation. In wet preparations examined under dark-field or phase-contrast microscopy, borreliae display a serpentine-like motility. In tissue sections, they are readily visible by silver (Dieterle, Warthin–Starry) stain.

Leptospira

Only one pathogenic species, *Leptospira interrogans*, exists but 200 or so serologic types have been identified on the basis of the hosts they infect, e.g. cats are host for serovar *icterohaemorrhagiae*, dogs for serovar *pomana*, and swine for serovar *canicola*. Humans are infected from contact of exposed skin with urine from these animal reservoirs, either directly or by urine contamination of soil, mud and surface waters. Cattle, raccoons, goats and mice are also reservoirs for human infection. In animal reservoirs, *Leptospira* cause bacteremia and abortion. In surviving animals, leptospires reside asympto-matically for years in the lumen of nephritic tubules, from where they are excreted in the urine. Leptospires may also be transmitted through milk,

placenta, and aborted fetuses. Other routes of acquisition are through mucous membranes of the oropharynx, conjunctiva, and genital tract. Leptospirosis is a zoonosis with world-wide distribution. Heavy tropical rains favor survival of leptospires. Leptospires enter skin and mucous membranes through breaks. Those with occupations involving contact with infected animals or their urine, such as agricultural workers, sewer workers, livestock handlers, military personnel, abattior workers, veterinarians and miners, are at increased risk. Human infections manifest as hemorrhage, diarrhea, jaundice, renal tubular necrosis, and hepatic dysfunction. Most manifestations result from damage to capillary endothelial linings, and tubular and hepatic dysfunction. Leptospires can be found in blood and in various organs and meninges. Severe cases of leptospirosis are characterized by diphasic manifestations, including an anicteric septicemia of 4–7 days followed by an icteric (jaundice) phase (Weil's disease), which can be fatal.

Morphology

Leptospira are flexible, helical rods with more than 18 right-handed helicals (coils) per cell. Usually, one or both ends of the cell are hooked. Leptospires are faintly Gram-negative and are best demonstrated by dark-field or phase-contrast microscopy. Leptospires are motile by rotation around their long axis.

Culture characteristics

Leptospires can be cultivated in Fletcher's or Stuart's semi-solid media enriched with rabbit serum or bovine serum albumin, in which they grow in the dark just below the surface.

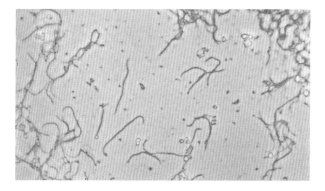

Figure 469 *Leptospira icterohaemorrhagiae* Phase-contrast-enhanced microscopy showing delicate, tightly coiled bacillus with spirals at regular intervals and with hooks at one or both termini

Treponema pallidum

Treponema pallidum is solely a human pathogen with infection spread from an infected to non-infected individual through sexual contact, by transplacental passage (congenital syphilis), and through blood transfusion. The microorganism gains access to tissues by penetration through intact mucosae or through abraded skin. Local replication takes place and the bacterium enters the lymphatics and is widely disseminated through the bloodstream to many organs and skin (secondary syphilis). At the site of entry a primary lesion (syphilitic chancre) develops after about 3 weeks' incubation. Skin lesions of secondary syphilis usually involve the entire trunk and extremities, including the palms of the hands and soles of the feet. Lesions are macular or pustular, teeming with treponemes, and are highly infectious. Lesions are not chancre-like because of incipient humoral and cellular immunity. Oral

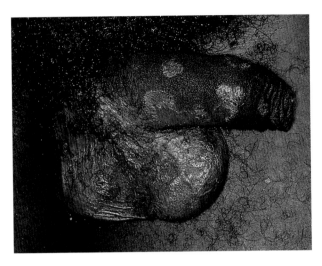

Figure 472   *Treponema pallidum* Secondary syphilis of penis manifested by dry, flat, non-purulent, plaque-like scaling lesions admixed with primary raised, oozing chancre on scrotum. In 15% of patients with secondary syphilis, the primary chancre may also be present

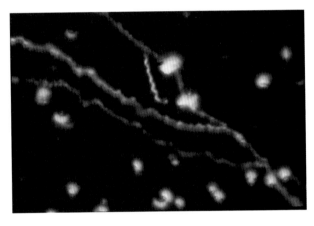

Figure 470   *Treponema pallidum* Thin, delicate spirochete with identical tapering ends in dark-field microscopy of fresh exudate from penile lesion

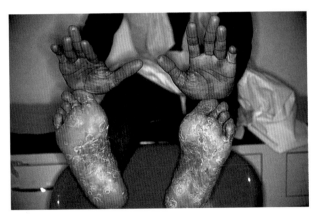

Figure 473   *Treponema pallidum* Flat, dry, scaling lesions of secondary syphilis on palms of hands and soles of feet

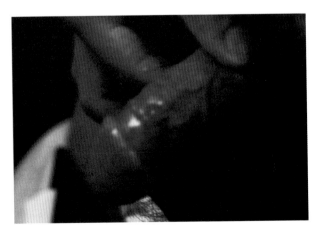

Figure 471   *Treponema pallidum* Syphilitic chancre with indurated margin on erythematous non-purulent base

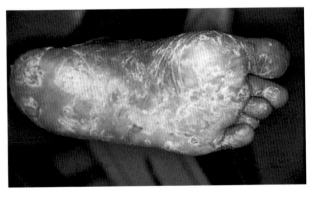

Figure 474   *Treponema pallidum* Higher magnification of flat, scaling lesions of secondary syphilis on sole of foot. Lesions, although dry, when unroofed were teeming with spirochetes and, hence, highly infectious

mucosal lesions and gastric lesions also occur. In the absence of antibiotic treatment, secondary disease will resolve and latent syphilis ensues which can remain dormant for years. Relapses during the first 4 years (early latency) of latency can occur. Although spirochetes may be difficult to detect during late latency, non-specific VDRL (Venereal Diseases Research Laboratory) and specific TPHA (*Treponema pallidum hemagglutination*) antibody tests remain positive. Tertiary syphilis develops after a variable latency, usually exceeding 2 years. It largely reflects manifestations of the immune response to affected tissues and encompasses gummas (granulomatous lesions of skin or bone), cardiovascular (aortitis), and neurosyphilis. Virulence factors of *T. pallidum* include hyaluronidase, which aids penetration of glycocalyx surrounding host cells, fibronectin-associated adherence to endothelial cells (syphilis is largely a vasculitis), and presence of glycosamino-glycan and sialic acid in its external layer, which inhibit bactericidal activity of the classic and alternative complement pathways. Treponemes can be propagated in laboratory animals. *Treponema carateum* is the etiologic agent of pinta, a non-sexually transmitted infection involving the skin only. *T. carateum* is not as contagious as the three subspecies of *T. pallidum*, but its skin manifestations result in disfigurement and social stigmata. Although the precise mode of *T. carateum* transmission is unknown, spread of infection is thought to occur by direct contact with open lesions which may be flat (pintids) after progression from an initial papule or plaque. Crops of pintids (symptomatic episodes) may reappear for 40 years, extending pinta infectivity. *T. pallidum* subspecies *pertenue* causes yaws, a chronic, relapsing, contagious skin infection, also transmitted by direct skin contact with the exudate from an infectious lesion; in both pinta and yaws, entry of the microorganism is facilitated by breaks in the skin, scabes infestation, etc. Yaws is the most prevalent endemic treponematosis affecting populations in rural areas of the world where high levels of humidity and rainfall prevail. The initial site of entry of the microorganism is on the legs, feet or buttocks, where a small, erythematous papule develops which enlarges into a papilloma. *Treponema pallidum* subspecies *endemicum* causes a non-venereal syphilis-like disease called endemic syphilis or bejel, which is prevalent in Africa, Western Asia, and Australia. Transmission of this 'pre-pubertal' syphilis is direct by person to person and by sharing contaminated feeding utensils.

The main reservoir for the microorganism is in children 2–15 years of age. The highest prevalence of infection is in dry, hot, temperate climates. Primary lesions may occur in the mouth, lips, and nipples of breast-feeding women. Analogous to venereal syphilis, secondary and tertiary lesions (gummata of nasopharynx, skin and bone) also occur.

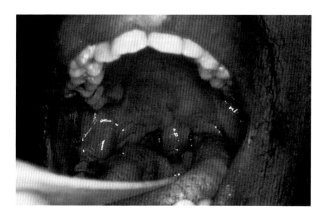

Figure 475   *Treponema pallidum*   Ulcerated mucosal lesion in patient with secondary syphilis

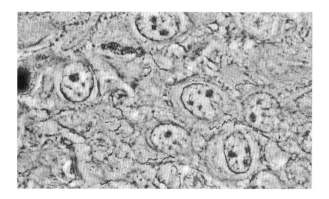

Figure 476   *Treponema pallidum* Numerous spirochetes in phase-contrast-enhanced microscopy of Dieterle silver stain of skin biopsy from patient with secondary syphilis

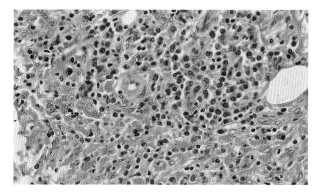

Figure 477   *Treponema pallidum*   Perivascular plasma cell infiltrate in hematoxylin and eosin stain of skin biopsy of patient with secondary syphilis

12

Intracellular bacterial species

CALYMMATOBACTERIUM GRANULOMATIS

C. granulomatis is a pleomorphic, *Klebsiella*-like, encapsulated coccobacillus occurring intracellular in histocytes. It causes 'granuloma inguinale', a destructive, ulcerative, mainly sexually transmitted disease. Unlike in chancroid, which it may mimic, regional lymphadenopathy is absent. Lesions are large, chronic, and may be misdiagnosed as a carcinoma. Metastatic spread of the microorganism via the bloodstream may result in lesions in bones, joints, and viscera. As *C. granulomatis* cannot be grown on bacteriologic media, diagnosis is achieved by visualizing intracellular encapsulated coccobacilli in Giemsa-stained smears of ulcer scrapings.

CHLAMYDIA

Chlamydiae are obligate intracellular pathogens, which are widely distributed in nature and cause ocular, genitourinary and pulmonary infections in humans. An association with atherosclerosis is under intense investigation. The family *Chlamydiaceae* previously contained three human pathogens, *C. trachomatis*, *C. pneumoniae* and *C. psittaci*. The latter two species have been removed into a new genus, *Chlamydophila*, on the basis of ribosomal sequence data. *C. trachomatis* is divided into three biovars, based upon their human disease potential: the trachoma biotype, which causes trachoma, the TRIC biotype which causes inclusion conjunctivitis, genital tract infections and afebrile pneumonitis in infants, and the lymphogranuloma venereum (LGV) biotype, which causes a more invasive genital tract infection. *C. psittaci* is widespread among avian hosts and causes a zoonotic pneumonia (psittacosis) in

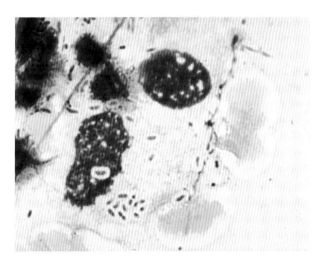

Figure 478 *Calymmatobacterium granulomatis* Intracellular and extracellular encapsulated diplobacilli ('Donovan bodies') in Giemsa-stained smear of scraping of penile ulcer

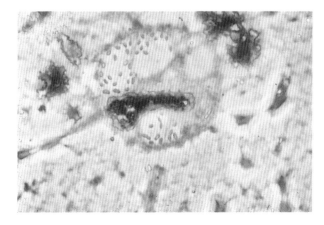

Figure 479 *Calymmatobacterium granulomatis* Phase-contrast-enhanced microscopy of Giemsa stain of scraping of penile ulcer, showing numerous diplobacilli within histiocyte

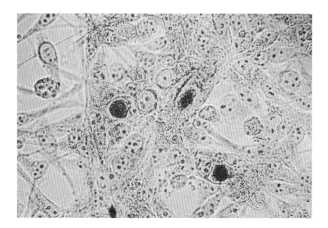

Figure 480 *Chlamydia trachomatis* Iodine-stained McCoy cell tissue culture showing dark brown glycogen-containing inclusion (reticulate body) containing maturing elementary bodies

humans. *C. pneumoniae* is associated with mild to acute respiratory disease in young adults. Chlamydiae are distinguished from other bacteria by virtue of their unique dimorphic growth cycle, consisting of an environmentally resistant, metabolically inert, extracellular infectious elementary body, and an intracellular replicating form, the reticulate body. Elementary bodies are small, rigid, ovoid to pear-shaped, 'spore-like' bodies, enabling extracellular survival. The chlamydial infection cycle begins by attachment of the elementary body to, and ingestion by, susceptible, non-phagocytic, non-ciliated columnar or cuboidal host epithelial cells, such as those found in the conjunctiva, endocervix, urethra, rectum, and mucosa of the endometrium and Fallopian tubes. Six to 8 hours after entering a host cell, in the setting of failure of cellular lysosomes to fuse with the elementary body-containing phagosome, the elementary body becomes reorganized into the metabolically active and dividing form, the non-infectious reticulate body. Using host cell precursors, the reticulate body synthesizes all necessary macromolecules (RNA, DNA, proteins) and divides by binary fission with some of the 'daughter cells', reorganizing into elementary bodies. As the phagosome enlarges, it assumes a distinct morphology (inclusion) and translocates to a perinuclear position within the infected cells. Between 48 and 72 h post-infection, the inclusion is extruded from the host cell, releasing infectious elementary bodies into the extracellular environment, with continued cell-to-cell and host-to-host transmission. *C. trachomatis* serovars cause ocular infections such as trachoma, inclusion conjunctivitis, ophthalmia neonatorum in infants, genital tract infections in males and females, and lymphogranuloma venereum, a multi-system infection. Complications of *Chlamydia* infection in women include endometritis and salpingitis (pelvic inflammatory disease), among other manifestations. Asymptomatic carriage of *C. trachomatis* is common in both males and females.

Morphology

In conjunctival scrapings in patients with trachoma, conjunctivitis, or infant inclusion disease (ophthalmia neonatorum), two types of intracytoplasmic inclusions may be visualized: elementary bodies and the reticulate perinuclear initial body.

Culture characteristics

Diagnosis of *C. trachomatis* infection is through inoculation of cell culture of cyclohexamide-treated McCoy cells and demonstration of the characteristic iodine-staining, glycogen-positive perinuclear reticulate inclusion body. Inclusions can also be demonstrated by Giemsa and immunofluorescence staining. For *C. psittaci* and *C. pneumoniae*, the inclusions can be visualized with genus-specific monoclonal antibodies or by Giemsa stain.

MYCOPLASMA AND UREAPLASMA

Mycoplasma and *Ureaplasma* are the smallest prokaryotes capable of growth on bacteriologic media. On agar substrates, they form colonies best visualized by direct microscopy of the agar surface. Mycoplasmas are widely distributed in nature, producing infections in livestock, birds, cold-blooded animals and humans. Ureaplasmas, originally designated 'T' (for tiny) strain *Mycoplasma* because of their small colony size, are also widely distributed among animals. Animal-derived strains are antigenically distinct from human isolates. *Mycoplasma* and *Ureaplasma* belong to the class Mollicutes ('soft skin') in reference to their marked pleomorphism and the absence of a cell wall containing peptidoglycan (which also accounts for their resistance to cell wall-acting antibiotics). The genus *Mycoplasma* contains more than 100 named species, of which *M. hominis*, *M. pneumoniae*, *M. genitalium* and *M. fermentans* are the more common human isolates. *U. urealyticum*, a urease-producing species, is one of six species in the genus *Ureaplasma*, but clinically the most relevant as a cause of urogenital tract infection. Mycoplasmas, which are bounded only by a plasma membrane, multiply by binary fission, sometimes asynchronously between genomic replication and cytoplasmic division, which results in marked pleomorphism and filament formation. Mycoplasmas require sterols for growth, usually provided by animal serum supplementation of media. Clinically, mycoplasmas have been associated with a variety of human infections of the respiratory and urogenital tracts. *M. pneumoniae* causes individual and sporadic outbreaks of pneumonia, with children being more frequently infected than young adults. Severe pneumonias can occur in immunodeficient hosts. *M. hominis* is an occasional incitant of pneumonia and meningitis in

newborns and has been recovered from the male and female genital tract, where it may have an inconclusive role in non-gonococcal urethritis (males) and pelvic inflammatory disease. A role for *M. hominis* and *U. urealyticum* in spontaneously aborted fetuses has been suggested, based upon the recovery of these mycoplasmas from amniotic fluid of women with chorioamnionitis and from aborted fetuses. Furthermore, both *M. hominis* and *U. urealyticum* have been recovered from the blood of some febrile women post-abortion. *M. hominis* has been reported to cause surgical infections. *M. fermentans* has been recovered from the peripheral blood monocytes of AIDS patients and was originally thought to contribute to the acceleration of the progession of AIDS, but this concept remains unproven.

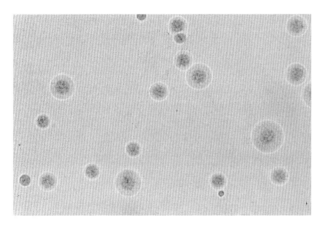

Figure 481 *Mycoplasma hominis* Flat 'fried egg' colonies on New York City agar with dense central core which is embedded in the agar substrate, and spreading peripheral zone of surface growth

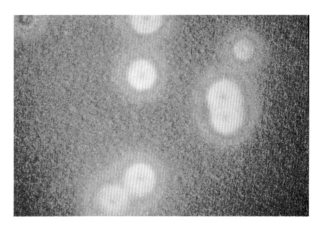

Figure 483 *Mycoplasma pneumoniae* Microscopic depiction of clear zones of β-hemolysis surrounding colonies on mycoplasma agar after flooding agar surface with guinea pig erythrocyte suspension. Erythrocytes adhere to colonies (hemadsorption) and then lyse by action of hemolysin (peroxide)

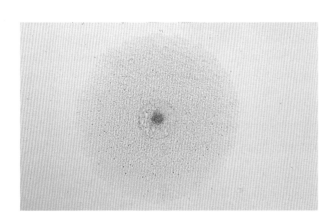

Figure 482 *Mycoplasma hominis* Higher magnification of 'fried egg' colony, showing finer details of dense central core surrounded by defined narrow zone, and granular zone of spreading surface growth

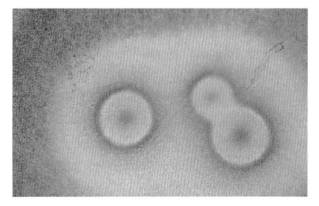

Figure 484 *Mycoplasma pneumoniae* High-power magnification of zone of complete hemolysis of guinea pig erythrocytes. Sheep erythrocytes are also lysed

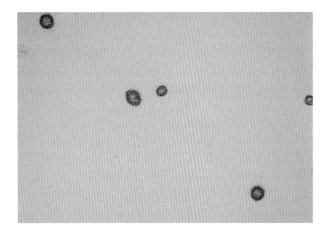

Figure 485 *Ureaplasma urealyticum* Dark golden-brown colonies with little peripheral growth on urea containing medium. Hydrolysis of urea yields ammonia, which oxidizes manganous chloride in medium to manganese dioxide, producing a colony-associated brownish precipitate

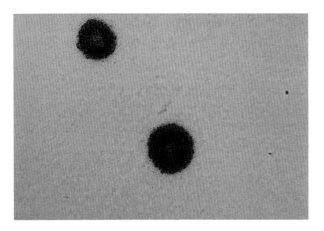

Figure 486 *Ureaplasma urealyticum* High-power magnification showing roughly textured golden-brown colonies

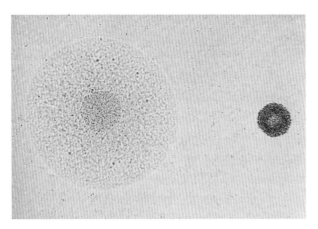

Figure 487 *Mycoplasma hominis* and *Ureaplasma urealyticum* Comparative size relationship between *M. hominis* colony (left) and *U. urealyticum* colony (right) on urea containing agar medium

Morphology

Mycoplasmas and Ureaplasmas do not stain well, as they lack a rigid cell wall for stain entrapment. Furthermore, their extensive plasticity does not allow for a definitive morphologic presentation.

Culture characteristics

M. hominis grows on most blood-enriched bacteriologic media, especially under anaerobic conditions after 48-h incubation. On such media, however, colonies are tiny and may be overlooked unless plates are carefully examined by oblique lighting. On blood-free enriched media such as A7 agar or New York City medium, *M. hominis* colonies are flat with a dense central core and radiating periphery of growth, rendering a typical 'fried egg' appearance. Colonies of *M. pneumoniae* are more compact, do not present with a 'fried egg' appearance, and can be identified by flooding the agar surface with guinea pig erythrocytes and observing for hemadsorption of the erythrocytes to *M. pneumoniae* colonies and subsequent hemolysis of the erythrocytes. *U. urealyticum* colonies developing on a urea-containing medium are small and dark golden-orange.

ANAPLASMA (EHRLICHIA) PHAGOCYTO-PHILA AND EHRLICHIA

Anaplasmataceae, in the order Rickettsiales, are *Rickettsia*-like, Gram-negative, obligately intracellular parasites of circulating leukocytes. *Rickettsia* differ from *Anaplasma* and *Ehrlichia* by multiplying in the cytoplasm of endothelial cells. Depending on the species, *Anaplasma* and *Ehrlichia* chiefly infect monocytes (*E. chaffeensis*), granulocytes (*A. phagocytophila*), lymphocytes, or platelets, and can be transmitted through blood transfusions, in addition to natural tick vectors. These bacteria are pleomorphic and complete their life cycle within cytoplasmic vacuoles of parasitized cells, where they undergo binary fission to form clusters of bacteria called morulae. Survival in phagocytic vacuoles is related to inhibition of phagosome–lysome fusion and inhibition of superoxide anion production. Their maturation processes in phagocytes resemble those of *Chlamydia* and include an elementary body (infectious form), initial body, and mature morula stages. Morulae break out of host cells and release elementary bodies, continuing the cycle. Part of the *Anaplasma* life cycle takes place in a hard-shell

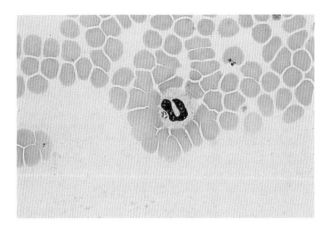

Figure 488 *Anaplasma phagocytophila* Slightly ovoid, lavender-staining morula containing four bacillary forms. The morula (phagocytic vacuole) is situated adjacent to the neutrophil nucleus

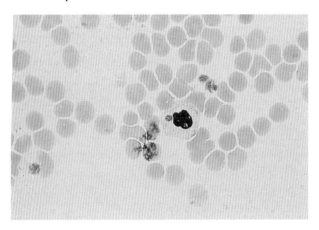

Figure 489 *Anaplasma phagocytophila* Discrete, dark, oval, lavender-staining morula containing numerous bacillary forms adjacent to neutrophil nucleus. Peripheral blood smear from 68-year-old febrile patient who recalled tick bite while hiking in Westerchester County, New York, an endemic area for Lyme disease and human granulocytic ehrlichiosis

arthropod host: *Ixodes scapularis* (deer tick) in the north eastern United States (which also transmits the agent of Lyme disease *Borrelia burgdorferi*, and *Babesia microti*, a malarial-like parasite of erythrocytes), and *Ixodes pacificus* in the Pacific western states. *E. chaffeensis* is transmitted by *Amblyomma americanum* ticks. As transovarial transmission of *Anaplasma* and *Ehrlichia* species is inefficient in ticks, mammalian human and animal (dogs, sheep, goats, cattle, horses) help to maintain *Anaplasma* in nature. *A. phagocytophila* is the agent of human granulocytic ehrlichiosis (HGE), a febrile illness accompanied by myalgias, headache, elevated serum liver transaminase, and leukopenia and thrombocytopenia in severe cases. Most infections occur between May and October when nymphal deer ticks are most active in search of a blood meal. Adult ticks quest during autumn months.

Morphology

Examination of Wright- or Giemsa-stained smears of peripheral blood (preferably buffy coats) may reveal morulae in the cytoplasm of neutrophils or other nucleated white blood cells. Morulae usually stain darker than the cell nucleus, are present in various sizes and number, and are stippled with bacillary forms, ranging from one to several.

Culture characteristics

A. phagocytophilia has been cultivated in HL60 cell culture, a line of human promyelocytic leukemic cells, in which cytophatic effects with lysis of cell take place within 5–12 days after inoculation with blood from an infected patient.

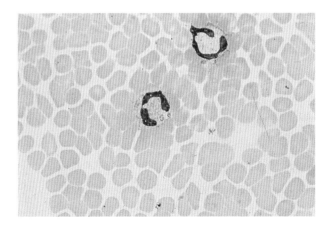

Figure 490 *Anaplasma phagocytophila* Neutrophil containing two morulae with discrete, lavender-staining bacillary forms

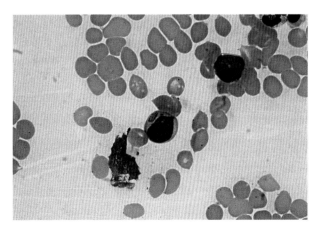

Figure 491 *Anaplasma phagocytophila* Crescent-shaped, lavender-staining morula with numerous bacillary forms in lymphocyte in peripheral blood smear

Bibliography

Gram-positive, aerobic, branching and non-branching, non-acid-fast bacillary species

Bottone EJ, Levine S, Burton R, *et al*. Unclassified *Bacillus* species resembling *Bacillus anthracis*: potential for misdiagnosis during anthrax alert. *Clin Microbiol Newsl* 2003;25:49–53

Bottone EJ, Sierra MF. *Listeria monocytogenes*: another look at the 'Cinderella' among pathogenic bacteria. *Mt Sinai J Med* 1977;44:42–59

Braun L, Cossart P. Interactions between *Listeria monocytogenes* and host mammalian cells. *Microbes and Infection* 2000;2:803–11

Brooke CJ, Riley TV. *Erysipelothrix rhusiopathiae*: bacteriology, epidemiology, and clinical manifestations of an occupational pathogen. *J Med Microbiol* 1999;48:789–99

Brown JM, Georg LK, Waters LC. Laboratory identification of *Rothia dentocariosa* and its occurrence in human clinical materials. *Appl Microbiol* 1969;17:150–6

Brown JR. Human actinomycosis. A study of 181 subjects. *Hum Pathol* 1973;4:319–30

Davey Jr RT, Tauber WB. Posttraumatic endophthalmitis: the emerging role of *Bacillus cereus* infection. *Rev Infect Dis* 1987;9:110–23

Dixon T, Meselson M, Guillemin J, *et al*. Anthrax. *N Engl J Med* 1999;341:815–26

Faoagali JL. *Kurthia*, an unusual isolate. *Am J Clin Pathol* 1974;62:604–6

Funke G, vonGraevenitz A, Clarridge JE, *et al*. Clinical microbiology of coryneform bacteria. *Clin Microbiol Rev* 1997;10:125–59

Georg LK, Brown JM. *Rothia*, gen. nov. An aerobic genus of the family *Actinomycetaceae*. *Int J Syst Bacteriol* 1967;1:79–88

Jordan HV, Kelly M, Heeley HD. Enhancement of experimental actinomycosis in mice by *Eikenella corrodens*. *Infect Immun* 1984;46:367–71

LeProwse CR, McNeil MM, McCarthy JM. Catheter-related bacteremia caused by *Oerskovia turbata*. *J Clin Microbiol* 1989;27:571–2

Lorber B. Listeriosis. *Clin Infect Dis* 1997;24:1–11

Pancoast SJ, Ellner PD, Jahre JA, *et al*. Endocarditis due to *Kurthia bessoni*. *Ann Intern Med* 1979;90:936–7

Pope J, Singer C, Kiehn TE, *et al*. Infective endocarditis caused by *Rothia dentocariosa*. *Ann Intern Med* 1979;91:746–7

Reboli AC, Farrar WE. *Erysipelothrix rhusiopathiae*: an occupational pathogen. *Clin Microbiol Rev* 1989;2:354–9

Reller, LB, Maddoux GL, Eckman MR, *et al*. Bacterial endocarditis caused by *Oerskovia turbata*. *Ann Intern Med* 1975;83:664–6

Smith R, Henderson P. Actinomycotic canaliculitis. *Aust J Ophthalmol* 1980;8:75–9

Southwick FS, Purich DL. Intracellular pathogenesis of listeriosis. *N Engl J Med* 1996;334:769–76

Tilney LG, Tilney MS. The wily ways of a parasite: induction of actin assembly by *Listeria*. *Trends Microbiol* 1993;1:25–31

Turnbull PCB. Definitive identification of *Bacillus anthracis* – a review. *J Appl Microbiol* 1999;37:237–40

Gram-positive, partially acid-fast, branched, aerobic species

McNeil MM, Brown JM. The medically important aerobic actinomycetes: epidemiology and microbiology. *Clin Microbiol Rev* 1994;7:357–417

Prescott JF. *Rhodococcus equi*: an animal and human pathogen. *Clin Microbiol Rev* 1991;4:20–34

Mycobacterium

Damsker B, Bottone EJ, Deligdisch L. *Mycobacterium xenopei*: infection in an immunosuppressed host. *Hum Pathol* 1982;13:866–9

Engler HD, Hass A, Hodes DS, *et al*. *Mycobacterium chelonei* infection of a Broviac catheter site. *Eur J Clin Microbiol* 1989;8:521–4

Wallace Jr RJ, Swenson JM, Silcox VA, *et al*. Spectrum of disease due to rapidly growing mycobacteria. *Rev Infect Dis* 1983;5:657–79

Wayne LG, Sramek HA. Agents of newly recognized or infrequently encountered mycobacterial diseases. *Clin Microbiol Rev* 1992;5:1–25

Gram-negative, fermentative, aerobic bacilli

Bottone EJ, Schneierson SS. *Erwinia* species: an emerging human pathogen. *Am J Clin Pathol* 1972;57:400–5

Bottone EJ. *Yersinia enterocolitica*: the charisma continues. *Clin Microbiol Rev* 1997;10:257–76

Chuang Y, Young C, Chen C. *Vibrio vulnificus* infection. *Scand J Infect Dis* 1989;21:721–6

Holmberg SD. *Vibrios* and *Aeromonas*. *Infect Dis Clin North Am* 1988;2:655–77

Janda JM, Abbott SL. *The Enterobacteria*. Philadelphia: Lippincott-Raven, 1998

Janda JM, Brendan R, DeBenedetti J, *et al*. *Vibrio alginolyticus* bacteremia in an immunocompromised patient. *Diagn Microbiol Infect Dis* 1986;5:337–40

Janda JM, Duffy PS. Mesophilic aeromonads in human disease: current taxonomy, laboratory identification, and infectious disease spectrum. *Rev Infect Dis* 1988;10:980–97

Kaysner CA, Abeyta Jr, C, Trost PA, *et al*. Urea hydrolysis can predict the potential pathogenicity of *Vibrio parahaemolyticus* strains isolated in the Pacific northwest. *Appl Environ Microbiol* 1994;60:3020–2

Namdari H, Bottone EJ. *Aeromonas* species: pathogens of aquatic inhabitants with a human host range. *Clin Microbiol Newsl* 1991;13:113–16

Ponte R, Jenkins S. *Fatal Chromobacterium violaceum* infections associated with exposure to stagnant water. *Pediatr Infect Dis* 1992;11:583–6

Sorensen RV, Jacobs MR, Shurin SB. *Chromobacterium violaceum* adenitis acquired in the northern United States as a complication of chronic granulomatous disease. *Pediatr Infect Dis* 1985;4:701–2

Stromm MS, Paranjpye RN. Epidemiology and pathogenesis of *Vibrio vulnificus*. *Microbes and Infect* 2000;2:177–88

Yoshida S, Ogawa M, Mizuguchi Y. Relation of capsular materials and colony opacity to virulence of *Vibrio vulnificus*. *Infect Immun* 1985;47:446–51

Non-fermenting, Gram-negative aerobic bacilli

Bottone EJ, Reitano M, Janda JM, *et al*. *Pseudomonas maltophilia* exoenzyme activity as a correlate in the pathogenesis of ecthyma gangrenosum. *J Clin Microbiol* 1986;24:995–7

Bouvet PJM, Grimont PAO. Taxonomy of the genus *Acinetobacter* with the recognition of *Acinetobacter baumanii* sp nov., *Acinetobacter haemolyticus* sp non., *Acinetobacter johnsonii* sp. nov., and *Acinetobacter junii* sp nov. and extended descriptions of *Acinetobacter calcoaceticus* and *Acinetobacter lwoffii*. *Int J Syst Bacteriol* 1986;36:228–40

Govan JRW, Deretic V. Microbial pathogenesis in cystic fibrosis: mucoid *Pseudomonas aeruginosa* and *Burkholderia cepacia*. *Microbiol Rev* 1996;60:539–74

Iglewski BH, Kabat D. NAD-dependent inhibition of protein synthesis by *Pseudomonas aeruginosa* toxin. *Proc Natl Acad Sci USA* 1975;72:2284–8

Pitt TL. Biology of *Pseudomonas aeruginosa* in relation to pulmonary infection in cystic fibrosis. *J Roy Soc Med* 1986;79(Suppl 12):13–18

Pollack M. *Pseudomonas aeruginosa* exotoxin A. *N Engl J Med* 1980;302:1360–2

Srinivasan A, Kraus CN, Deshazer D, *et al*. Glanders in a military research microbiologist. *N Engl J Med* 2001;345:256–8

Anaerobes

Bartlett JG. *Clostridium difficile*: history of its role as an enteric pathogen and the current state of knowledge about the organism. *Clin Infect Dis* 1994;18(Suppl 4):S265–72

Bottone EJ, Lee P. Fusospirochetal superinfection of pre-existing oral lesions in patients with acquired immunodeficiency syndrome. *Diagn Microbiol Infect Dis* 1997;28:51–3

Freeman J, Wilcox MH. Antibiotics and *Clostridium difficile*. *Microbes Infect* 1999;1:377–84

Kristensen LH, Prag J. Human necrobacillosis, with emphasis on Lemierre's syndrome. *Clin Infect Dis* 2000;31:524–32

Narula A, Khatib R. Characteristic manifestations of *Clostridium* induced spontaneous gangrenous myositis. *Scand J Infect Dis* 1985;17:291–4

Stevens DL, Musher DM, Watson DA, *et al*. Spontaneous, nontraumatic gangrene due to *Clostridium septicum*. *Rev Infect Dis* 1990;12:286–96

Fastidious, Gram-negative bacilli

Bottone EJ, Granato PA. *Eikenella corrodens* and closely related bacteria. In Dworkin M, Falkow S, Rosenberg E, *et al*. eds. *The Prokaryotes*: *An Evolving Electronic Resource for the Microbiological Community*, 3rd edn. New York: Springer Verlag, 2000

Morrison VA, Wagner KW. Clinical manifestations of *Kingella kingae* infections: case report and review. *Rev Infect Dis* 1989;5:776–82

Meyers BR, Bottone EJ, Lasser R, *et al*. Infection due to *Actinobacillus actinomycetemcomitans*. *Am J Clin Pathol* 1971;56:204–11

Wormser GP, Bottone EJ. *Cardiobacterium hominis*: review of microbiologic and clinical features. *Rev Infect Dis* 1983;5:680–91

Bottone EJ, Allerhand J. Association of mucoid encapsulated *Moraxella duplex var. nonliquefaciens* with chronic bronchitis. *Appl Microbiol* 1968;16:315–19

Brenner DJ, Hollis DG, Fanning GR, *et al*. *Capnocytophaga canimorsus* sp. nov. (formerly CDC Group DF-2), a cause of septicemia following dog bite, and *C cynodegmi* sp. nov., a cause of localized wound infections following dog bite. *J Clin Microbiol* 1989;27:231–5

Catlin BW. *Gardnerella vaginalis*: characteristics, clinical considerations, and controversies. *Clin Microbiol Rev* 1992;5:213–37

Guthrie R, Bakenhaster K, Nelson R, *et al*. *Branhamella catarrhalis* sepsis: a case report and review of the literature. *J Infect Dis* 1988;158:907–8

Huprikar S, Bottone EJ. *Francisella* and *Pasteurella* and *Yersinia pestis*. In Gorbach SL, Bartlett JG, Blacklow NR, eds. *Infectious Diseases*, 3rd edn. Orlando, Florida: WB Saunders, 2002

Karem KL, Paddock CD, Regnery RL. *Bartonella henselae, B. quintana*, and *B. bacilliformis*; historical pathogens of emerging significance. *Microb Infect* 2000;2:193–205

Morse SA. Chancroid and *Haemophilus ducreyi*. *Clin Microbiol Rev* 1989;2:137–57

Preston A, Maskell DJ. A new era of research into *Bordetella pertussis* pathogenesis. *J Infect* 2002;44:1–16

Qureshi MN, Lederman J, Neibart E, *et al. Bordetella bronchiseptica* recurrent bacteremia in the setting of a patient with AIDS and indwelling Broviac catheter. *Int J STD & AIDS* 1992;3:291–3

Schmidt H, Hansen JG. Diagnosis of bacterial vaginosis by wet mount identification of bacterial morphotypes in vaginal fluid. *Int J STD & AIDS* 2000;11:150–5

Trees DL, Morse SA. Chancroid and *Haemophilus ducreyi*: an update. *Clin Microbiol Rev* 1995;8:357–75

Van Langenhove G, Daelemans R, Zachee P, *et al. Pasteurella multocida* as a rare cause of peritonitis in peritoneal dialysis. *Nephron* 2000;85:283–4

von Konig CH, Halperin S, Riffelmann M, *et al.* Pertussis in adults and infants. *Lancet Infect Dis* 2002;2:744–50

Wolfrey BF, Moody JA. Human infections associated with *Bordetella bronchiseptica*. *Clin Microbiol Rev* 1991;4:243–55

Gram-positive, catalase-positive cocci

Sheagren JN. *Staphylococcus aureus*. The persistent pathogen. *N Engl J Med* 1984;310:1368–73, and 310:1437–42

Gram-positive, catalase-negative cocci

Baker CJ, Edwards MS. Group B streptococcal infections. In Remington JS, Klein JO, eds. *Infectious Diseases of the Fetus and Newborn Infant*, 4th edn. Philadelphia: WB Saunders, 1995:980–1054

Bisno A, Stevens DL. Streptococcal infections of skin and soft tissues. *N Engl J Med* 1996;334:240–5

Bottone EJ, Thomas CA, Lindquist D, *et al.* Difficulties encountered in identification of a nutritionally-deficient *Streptococcus* on the basis of its failure to revert to streptococcal morphology. *J Clin Microbiol* 1995;33:1022–4

Bottone EJ. Encapsulated *Enterococcus faecalis*: role of encapsulation in persistence in mouse peritoneum in absence of mouse lethality. *Diagn Microbiol Infect Dis* 1999;33:65–8

Bouvet A, Grimont F, Grimont PAD. *Streptococcus defectivus* sp. nov. and *Streptococcus adjacens* sp. nov., nutritionally variant streptococci from human clinical specimens. *Int J Syst Bacteriol* 1989;39:290–4

Catto BA, Jacobs MR, Shales DM. *Streptococcus mitis*. A cause of serious infections in adults. *Arch Intern Med* 1987;147:885–8

Christensen JJ, Vibits K, Ursing J, *et al. Aerococcus*-like organisms, a newly recognized potential urinary tract pathogen. *J Clin Microbiol* 1991;29:1049–53

Clark RB, Gordon RE, Bottone EJ, *et al.* Morphological aberrations of nutritionally deficient streptococci: association with pyridoxal (vitamin B_6) concentration and potential role in antibiotic resistance. *Infect Immun* 1983;42:414–17

Cunningham MW. Pathogenesis of group A streptococcal infections. *Clin Microbiol Rev* 2000;13:470–511

Garvie EI. Separation of the *genus Leuconostoc* and differentiation of leuconostocs from other lactic acid bacteria. In Bergman T, ed. *Methods in Microbiology*. New York: Academic Press, 1984;16:14–78

Handwerger S, Horowitz H, Coburn K, *et al.* Infection due to *Leuconostoc* species: six cases and review. *Rev Infect Dis* 1990;12:602–10

Hardy S, Ruoff KC, Catlin EA, *et al.* Catheter-associated infection with a vancomycin-resistant gram-positive coccus of the *Leuconostoc* species. *Pediatr Infect Dis* 1988;7:519–20

Jett BD, Huycke HW, Gilmore MS. Virulence of enterococci. *Clin Microbiol Rev* 1994;7:462–78

Mastro TD, Spika JS, Lozano P, *et al. Pediococcus acidilacti*: nine cases of bacteremia. *J Infect Dis* 1990;161:956–60

McCarthy LR, Bottone EJ. Bacteremia and endocarditis caused by satelliting streptococci. *Am J Clin Pathol* 1974;61:585–91

Murray BE. The life and times of the enterococcus. *Clin Microbiol Rev* 1990;3:46–65

Park JW, Grossman O. *Aerococcus viridans* infection. *Clin Pediatr* 1990;29:525

Riedel WJ, Washington JA. Clinical and microbiological characteristics of Pediococci. *J Clin Microbiol* 1990;28:1348–55

Ruoff K. Nutritionally variant streptococci. *Clin Microbiol Rev* 1991;4:184–90

Williams REO, Hirch A, Cowan ST. *Aerococcus*, a new bacterial genus. *J Gen Microbiol* 1953;8:475–80

Zierdt CH. Light microscopic morphology, ultrastructure, culture, and relationship to disease of the nutritional and cell-wall-deficient β-hemolytic streptococci. *Diagn Microbiol Infect Dis* 1992;15:185–94

Gram-negative cocci

Barquet N, Gasser I, Domingo P, *et al.* Primary meningococcal conjunctivitis: report of 21 patients and a review. *Rev Infect Dis* 1990;12:838–47

Bratigan B, Cohen MS, Sparling PF. Gonococcal infection: a model of molecular pathogenesis. *N Engl J Med* 1985;312:1683–94

Devoe IW. The meningococcus and mechanisms of pathogenicity. *Microbiol Rev* 1982;46:162–90

Duerden BI, ed. Meningococcal infection. *J Med Microbiol* 1988;26:161–87

Hook III EI, Holmes KK. Gonococcal infections. *Ann Intern Med* 1985;102:229–43

Periappuram M, Taylor MRH, Keane CT. Rapid detection of meningococci from petechiae in acute meningococcal infection. *J Infect* 1995;11:201–3

Raucher HS, Newton MJ, Stern RH. Ophthalmia neonatorum caused by a penicillinase-producing *Neisseria gonorrhoeae. J Pediatr* 1982;100:925–6

Young EJ, Cardella TA. Meningococcemia diagnosed by peripheral blood smear. *J Am Med Assoc* 1988;260:992

Curved and spiral species

Catrenich CE, Makin KM. Characterization of the morphologic conversion of *Helicobacter pylori* from bacillary to coccoid forms. *Scand J Gastroenterol* 1991;26 (Suppl 181):58–64

Halter F, Hurlimann S, Inauen W. Pathophysiology and clinical relevance of *Helicobacter pylori. Yale J Biol Med* 1992;65:625–38

Rosh JR, Kurfist LA, Benkov KJ, *et al. Helicobacter pylori* and gastric lymphonodular hyperplasia in children. *Am J Gastroenterol* 1992;87:135–9

Spirochetes

Antal GM, Lukehart SA, Meheus AZ. The endemic treponematosis. *Microb Infect* 2000;4:83–94

Johnson RC, Ritzi DM, Livermore BP. Outer envelope of virulent *Treponema pallidum. Infect Immun* 1973;8:291–5

Shapiro ED, Gerber MA. Lyme disease. *Clin Infect Dis* 2000;31:533–42

Anaplasma (Ehrlichia) phagocytophila

Aguero-Rosenfeld ME. Laboratory aspects of tick-borne diseases: Lyme, human granulocytic ehrlichiosis and babesiosis. *Mt Sinai J Med* 2003;70:197–206

Bakken JS, Krueth J, Wilson-Nordskog C, *et al.* Clinical and laboratory characteristics of human granulocytic ehrlichiosis. *J Am Med Assoc* 1996;275:195–205

Bakken JS, Dumler JS. Human granulocytic ehrlichiosis. *Clin Infect Dis* 2000;31:554–60

Dumler JS, Barbet AF, Bekker CP, *et al*. Reorganization of genera in the families Rickettsiaceae and Anaplasmataceae in the order Rickettsiales: unification of some species of *Ehrlichia* with *Anaplasma*, *Cowdria* with *Ehrlichia* and *Ehrlichia* with *Neorickettsia*, descriptions of six new species combinations and designation of *Ehrlichia equi* and 'HGE agent' as objective synonyms of *Ehrlichia phagocytophila*. *Int J Syst Evol Microbiol* 2001;51:2145–65

Goodman JL, Nelson C, Vitale B, *et al*. Cultivation of the causative agent of granulocytic ehrlichiosis. *N Engl J Med* 1996;334:209–15

Mott J, Rikihisa Y. Human granulocytic ehrlichiosis agent inhibits superoxide anion generation by human neutrophils. *Infect Immun* 2000;68:6697–703

Rikihisa Y. The tribe ehrlichia and ehrlichial diseases. *Clin Microbiol Rev* 1991;4:286–308

Chlamydia

Batteiger BE, Jones RB. Chlamydial infections. *Infect Dis Clin North Am* 1987;1:55–81

Jendro MC, Deutsch T, Korber B, *et al*. Infection of human monocyte-derived macrophages with *Chlamydia*

trachomatis induces apoptosis of T cells: a potential mechanism for persistent infection. *Infect Immun* 2000;68:6704–11

Wong Y, Gallagher PJ, Ward ME. *Chlamydia pneumoniae* and atherosclerosis. *Heart* 1999;222–38

Mycoplasma and ureaplasma

Baseman JB, Quackenbush RL. Preliminary assessment of AIDS-associated *Mycoplasma*. *ASM News* 1990;56:319–23

Cassell GH, Cole BC. Mycoplasmas as agents of human disease. *N Engl J Med* 1981;304:80–99

Granato PA, Poe L, Weiner LB. Use of modified New York City medium for growth of *Mycoplasma pneumoniae*. *Am J Clin Pathol* 1980;73:702–5

Pasquelle AW. Recognition of *Mycoplasma hominis* in routine bacteriology specimens. *Clin Microbiol Newsl* 1988;10:145–52

Taylor-Robinson D. Infections due to *Mycoplasma and Ureaplasma*: an update. *Clin Infect Dis* 1996:23:671–84

Index